TECHNICAL WRITING A–Z

A Commonsense Guide to Engineering Reports and Theses

Trevor M. Young

Library of Congress Cataloging-in-Publication Data

Young, Trevor M.
Technical writing A-Z : a commonsense guide to engineering
reports and theses / Trevor M. Young.
p. cm.
ISBN 0-7918-0236-1
1. Technical writing. I. Title.

T11.Y69 2005
808'.0666--dc22

2005041188

Foreword

Few people I know read the instructions before attempting to assemble a child's toy, or look at the *Read me* files that come with new software; these items are referred to when the person gets stuck. So I do not expect you to read this book from cover to cover before starting to write your report or thesis. *Technical Writing A–Z* is written in plain English; entries are in alphabetical order, with numerous links to other entries. First read the section called **Structure of reports and theses**; thereafter, dip into it when you need to know something. If you do not find what you are looking for within about three minutes, it is probably not here, and you will need to look elsewhere. (I have included a list of style manuals—covering a range of academic disciplines and publication formats—which can be used as a starting point.) Unless a distinction is made, the word *report* has been used for both reports and theses.

Please send me your comments on this book; suggestions for future editions would be most welcome (e-mail: technicalwriting@asme.org).

Trevor Young

Contents

A

Abbreviated terms

- Abbreviations (shortened forms of words), acronyms (words formed from the initial letters of other words), and initialisms (groups of letters, which are pronounced separately, taken from the initial letters of other words) usually require an explanation. The general rule is to spell out the full name when the item is first mentioned, followed by the abbreviation in parentheses—for example: *Onboard Oxygen Generating Systems* (OBOGS)—and thereafter to use the abbreviation. You can underline the letters that make up the abbreviation if you wish—for example: *ANOVA* (*Analysis of variance*).
- When the abbreviated term is itself widely understood (e.g., AIDS, DNA, NATO), it is not necessary to give the full name in the text; however, it could still be included in the list of abbreviations for completeness (DNA can also mean Distributed interNet Application).
- It is common practice not to use periods with such abbreviations (Internet Protocol, for example, would be written as *IP* rather than *I.P.*). Plurals are formed by adding an *s* without an apostrophe.
- The choice of the preceding indefinite article—*a* or *an*—is based on the pronunciation of the acronym or initialism (e.g., a CD, a DVD, a NASA report, an I/O card, an NT system, an SEM image [pronounced es-ee-em]).
- Be careful when using abbreviated terms as adjectives, as the last word making up the acronym or initialism can be inadvertently repeated (e.g., RAM memory, LCD display, PIN number).
- Do not expect readers to remember every abbreviation that you have defined, so make it easy for them and provide a list of abbreviated terms (see **Nomenclature**). This is placed after the **table of contents** or alternatively—although this is not common—at the start or end of the **appendix**.

❑ See also **Abbreviations (common)**, **Characteristic numbers**, **Chemical elements, compounds, and symbols** and **Units**.

Abbreviations (common)

A list of the commonly encountered general-use and scholarly abbreviations is given below. It is not necessary to define them in the report, as they are widely understood. As a general rule you should avoid using these abbreviations in passages of continuous text, but in tables, figures, lists, footnotes, endnotes, and so on, where space is often limited, their use is acceptable. Similarly, general-use abbreviations such as *chap.*, *e.g.*, *etc.*, *fig.*, and *i.e.* are usually restricted to text within parentheses. Abbreviations specifically used for referencing (for example: anon., et al., n.d., trans.) are not normally used for other purposes in reports.

Abbreviations: general-use and scholarly

Abbreviation	Meaning	Notes
ab init.	from the beginning	from Latin *ab initio*
abr.	abridged	
add.	addendum	
ad inf.	to infinity	from Latin *ad infinitum*
ad init.	at the beginning	from Latin *ad initium*
ad loc.	at the place	from Latin *ad locum*
a.k.a.	also known as	
Am.E.	American English	
anon.	anonymous	
app.	appendix	
art.	article	plural: arts.
bk.	book	
Br.E.	British English	
ca. *or* c.	about, approximately	from Latin *circa*
cf.	compare	from Latin *confer*
chap. *or* ch.	chapter	plural: chaps.
col.	column	plural: cols.
dept.	department	plural: depts.
diss.	dissertation	

Abbreviations: general-use and scholarly (*continued*)

Abbreviation	Meaning	Notes
div.	division	plural: divs.
do.	ditto (the same)	
ea.	each	
ed.	edited by, editor, edition	plural: eds.
e.g.	for example	from Latin *exempli gratia*
eng. *or* engin.	engineering	
eq.	equation	plural: eqs.
esp.	especially	
et al.	and others	from Latin *et alii*
etc.	and so forth	from Latin *et cetera*
et seq.	and the following	from Latin *et sequens*
ex.	example	plural: exs., exx.
fig.	figure	plural: figs.
fn.	footnote	
gen.	genus	
ibid. *or* ib.	in the same source	from Latin *ibidem*, used to refer to a previously cited work
id.	the same	from Latin *idem*
i.e.	that is to say	from Latin *id est*
incl.	inclusive	
inst.	instant	refers to the current month
l.	line	
loc. cit.	in the place cited	from Latin *loco citato*
max.	maximum	
min.	minimum	
misc.	miscellaneous	
n.	note	
N.B.	take careful note	from Latin *nota bene*
n.d.	no date	
neg.	negative	
no. *or* nr.	number	plural: nos. *or* nrs.
non seq.	it does not follow	from Latin *non sequitur*
n.p.	no place, no publisher	
obs.	obsolete	
op. cit.	in the work already cited	from Latin *opere citato*
p.	page	plural: pp.
par.	paragraph	plural: pars.

Abbreviations: general-use and scholarly (*continued*)

Abbreviation	Meaning	Notes
pl.	plate, plural	plural: pls. (plates)
pro tem.	for the time being	from Latin *pro tempore*
P.S.	postscript	from Latin *postscriptum*
pt.	part	plural: pts.
pub.	publisher, published, publication	plural: pubs.
Q.E.D.	which was to be demonstrated	from Latin *quod erat demonstrandum*
q.v.	which see	from Latin *quod vide*, instructs the reader to look elsewhere
repr.	reprint, reprinted	
rev.	review, revision, revised	
sc. *or* scil	namely	from Latin *scilicet*
sec.	section	plural: secs.
ser.	series	
sing.	singular	
sum.	summary	
suppl.	supplement	
s.v.	under the word	from Latin *sub voce*
trans. *or* tr.	translator, translation	
ult.	previous [month]	from Latin *ultimo* [*mense*]
ult.	ultimate	
univ.	university	
v.	see	from Latin *vide*, instructs the reader to look elsewhere
v. *or* vs.	versus, against	
v.i.	see below	from Latin *vide infra*
viz.	namely	from Latin *videlicet*
vol.	volume	plural: vols.
v.s.	see above	from Latin *vide supra*
w.r.t.	with respect to	
yr.	year	

Notes:

1 There are some Latin-based abbreviations that are widely understood (for example: ca., e.g., etc., i.e., N.B., viz.) and a few that are popular in academic writing (for example: et al., Q.E.D.), but there are an increasing

number that are now considered old fashioned (or even an indication that the writer is showing off). They would include *ab init.*, *ad inf.*, *ad init.*, *ad loc.*, *cf.*, *et seq.*, *ibid.*, *id.*, *non seq.*, *op. cit.*, *pro tem.*, *q.v.*, *sc.*, *s.v.*, *v.i.*, and *v.s.* (it is not recommended that these abbreviations be used).

2 Latin-based abbreviations such as *ca.* and *et al.* are often written in italics (as is the custom for foreign-language words); however, there is an increasing trend not to do so, as the abbreviations have effectively been adopted into English.

3 The abbreviations can have initial capital letters—for example, when used at the start of a sentence (which is rarely done in reports) or when used to identify an item in a document (e.g., Fig. 3.5, Chap. 2).

4 Periods are frequently omitted in modern writing for multiword abbreviations (e.g., NB, QED).

5 See also **Latinisms**.

Abbreviations: date and time

- Days of the week and months should be spelled out in the text, but in notes, tables, figures, references, and so on, abbreviations can be used (e.g., Sun., Mon., Tue. or Tues., Wed., Thu. or Thurs., Fri., Sat., Jan., Feb., Mar., Apr., Jun., Jul., Aug., Sep. or Sept., Oct., Nov., Dec.). Similarly, the informal abbreviation used for a particular year (e.g., *'98* for *1998*) should not be used in the text. Decades are written as numerals without an apostrophe (e.g., 1970s).

- As the date 2/8/99 can imply either *February 8, 1999* (American system) or *2 August 1999* (European system), the best approach is to avoid using an all-numeric date format and to spell out the month. Ordinal numbers (e.g., first, 1st, tenth) are not popular in American English, but are used in British English (e.g., 19th February 2005). An alternative format that has been adopted for much computer work is the ISO standard (8601:1988), which specifies the year-month-day, in that sequence, by ten keystrokes (e.g., 2003-01-18).

- Use numerals to indicate time—for example: write *9:00 A.M.* when using the 12-hour system and *15:20* (or *15h20*) for the 24-hour system, rather than *nine o'clock* and *twenty minutes past three*. Add *noon* or *midnight* after *12:00* to avoid miscommunication, unless it is absolutely clear that the 24-hour system is being used. There are a number of alternative formats that are also acceptable—for example: the abbreviations A.M. and P.M. are frequently written in lowercase (with or without periods), and a period is sometimes used in place of the colon.

- The ISO standard 8601:1988 recommends the 24-hour system, written in the sequence hours-minutes-seconds. The three elements are traditionally separated by colons—hours and minutes are each represented by two-digit values and seconds are expressed as decimals (e.g., *06:10:09.7* represents *6 hours*, *10 minutes*, *9.7 seconds*).
- The abbreviations for time zones are capitalized and written without periods (e.g., 17:30 EST, 19:00 UTC). The abbreviation UTC (Coordinated Universal Time) replaces GMT (Greenwich Mean Time), which should not be used.

Abbreviations: date and time

Abbreviation	Meaning	Example
A.D.	*anno Domini* (in the year of [our] Lord)	A.D. 943
B.C.	before Christ	137 B.C.
B.C.E.	before common era (equivalent to B.C.)	137 B.C.E.
B.P.	before present	4500 B.P.
C.E.	common era (equivalent to A.D.)	943 C.E.
A.M. *or* a.m.	*ante meridiem* (before noon)	11:23 A.M.
P.M. *or* p.m.	*post meridiem* (after noon)	10:00 P.M.

Note: The periods are frequently omitted in modern writing.

Abbreviations: titles and degrees

Social titles (such as Mr., Messrs., Mrs., Dr., M., MM., Mme.) are always abbreviated when preceding a name. Academic degrees (listed in the table that follows) and professional and honorary designations follow the name (e.g., Stephanie Swott, PhD; Martin Skywalker, FRAeS; J. J. Lawless, Sen.). In modern writing, social titles and abbreviations of this type are frequently written without periods.

Abbreviations: academic qualifications (not a complete list)

Abbreviation	Degree
A.B.	Artium Baccalaureus (Bachelor of Arts)
A.M.	Artium Magister (Master of Arts)
B.A.	Bachelor of Arts
B.A.A.S	Bachelor of Applied Arts and Sciences
B.D.	Bachelor of Divinity
B.Eng.	Bachelor of Engineering
B.F.A.	Bachelor of Fine Arts
B.M.	Bachelor of Medicine
B.Pharm.	Bachelor of Pharmacy
B.S. *or* B.Sc.	Bachelor of Science
B.V.Sc.	Bachelor of Veterinary Science
Ch.B.	Chirurgiae Baccalaureus (Bachelor of Surgery)
D.B.	Divinitatis Baccalaureus (Bachelor of Divinity)
D.D.	Divinitatis Doctor (Doctor of Divinity)
D.D.S.	Doctor of Dental Surgery
D.O.	Doctor of Osteopathy
D.Phil.	Doctor of Philosophy
D.S. *or* D.Sc.	Doctor of Science
D.V.M.	Doctor of Veterinary Medicine
J.D.	Juris Doctor (Doctor of Law)
L.H.D.	Litterarum Humaniorum Doctor (Doctor of Humanities)
Litt.D.	Litterarum Doctor (Doctor of Letters)
LL.B.	Legum Baccalaureus (Bachelor of Laws)
LL.D.	Legum Doctor (Doctor of Laws)
M.A.	Master of Arts
M.D.	Medicinae Doctor (Doctor of Medicine)
M.F.A.	Master of Fine Arts
M.Phil.	Master of Philosophy
M.S. *or* M.Sc.	Magister Scientia (Master of Science)
Ph.B.	Philosophiae Baccalaureus (Bachelor of Philosophy)
Ph.D.	Philosophiae Doctor (Doctor of Philosophy)
Ph.G.	Graduate in Pharmacy
S.B.	Scientiae Baccalaureus (Bachelor of Science)
S.M.	Scientiae Magister (Master of Science)

Note: The periods are frequently omitted in modern writing.

Abbreviations: use of periods

There is a definite trend to make writing less fussy and the dropping of periods in abbreviations is part of that trend. Many newspapers and magazines have almost eliminated the use of periods in abbreviations. In academic and professional documents, the period is retained in a number of situations. The following are guidelines:

- Use periods with
  - Latin abbreviations (such as: ca., e.g., etc., i.e., viz.);
  - Parts of a document (such as: app., chap., fig., p., par.);
  - Abbreviations used for referencing (such as: anon., ed., et al., n.d., pub., trans., vol.); or
  - Initials of people's names (e.g., P. L. Goodfellow).

- Do not use periods with
  - Acronyms and initialisms—this applies to well known terms (e.g., RAM, GDP) and to abbreviated terms defined in the report, for example: FMEA (Failure Mode and Effect Analysis);
  - Units of measurement (however, it may be necessary to add a period after *in*—for inches—to prevent a potential misunderstanding); or
  - Abbreviations of states in an address or citation reference (e.g., CA, MA, WA).

- For the rest, it is a matter of personal preference: abbreviations used with dates and time (e.g., BCE, AM), abbreviations of counties (UK, USA), social titles (e.g., Mrs, Dr), academic qualifications (e.g., BEng, PhD), professional or honorary designations (e.g., Sen, FRAeS), and so forth may be written with or without periods. It is important to be consistent, though, in the report.

Abbreviations (of journals)

- To save space in reference lists—and this is popular in journals, for example—the titles of cited journals are abbreviated. Similarly, the volume and page numbers of cited journal papers are also written in a condensed format—for example: "*J. Eng. Mat. & Tech.* 126(2004), 345–349" stands for "*Journal of Engineering Materials and Technology* volume 126 of 2004, pages 345 to 349." (Note that the house style of many journals is to set the volume number in bold. The journal title is always in italics.) However, in a report or thesis there is no real advantage in using an abbreviated title; in fact, using the full title may make it easier for readers to get copies of the cited work (journals may have similar titles).
- If you opt for abbreviating titles, use accepted abbreviations: do not make them up. Note that there is no single standard for abbreviating titles: different abstracting and indexing services use different methods (the journal mentioned earlier is also shortened to *J. Eng. Mater. Technol.*).
- A useful source for checking abbreviations is *Periodical Title Abbreviations* (Gale Research Co., Detroit, MI), which is compiled by abbreviation and by title. Reference details concerning journals and their abbreviations are also available at numerous university library websites.

Abstract

- An abstract is a descriptive summary of the work undertaken, without any figures or tables. It is written in the third person, predominantly in the past tense, although statements of opinion, conclusions, recommendations, and established facts may be presented in the present tense.
- It appears on its own page at the start of the report, under the heading *abstract*. It is usually between 200 and 400 words, and should never exceed one page. Short abstracts are presented as a single paragraph.

- The abstract should provide the readers with a concise overview of
  - (1) Why you did the work;
  - (2) What you did and how you did it; and
  - (3) The main results and conclusions.
- It should stand alone and be understandable independently of the rest of the report. If possible, avoid citing other publications, and never refer to chapters, figures, or tables contained within the report.
- The abstract is best written last—or, at least, after the substantive part of the report is finished. It is a good idea to write a longer draft of the abstract (say twice as long) covering all aspects of the work; and then to review it a few days later, consolidating and reducing the text until you feel that you have described all essential elements using the least number of words.
- As a check, write out a list of 10 to 15 keywords that describe your work and then see if they all appear in the abstract; revise it accordingly. Abstracts of reports, theses, and conference and journal papers are routinely indexed and stored on databases, enabling researchers to search for, and retrieve, abstracts containing keywords appropriate to their work.
- Typical structure:
  - Introductory sentence or two placing the work in context (mention previous work).
  - Brief statement on the objectives of the work.
  - Outline of the methodology and tools used (analytical or experimental).
  - One or two sentences stating the most important conclusions and/or recommendations.
- The abstract is one part of the report that is always read, so do not rush it and do it properly. See also **Summary** and **Executive summary**.

Acknowledgements

- It is polite to thank the kind individuals who, for example, proofread your report or set up your experimental apparatus. There is also no harm in flattering your supervisor or professor a little and including him or her in the acknowledgements for that *valuable support and expert knowledge*. Just do not go overboard and, for example, thank *Homer Simpson for all the laughs during the ordeal*. Such statements may appear to be a good idea at 3:00 a.m. (a few hours before the submission deadline), but when scrutinized in sober daylight they look silly.
- To do it right, keep your statements of gratitude simple and to the point. Use the person's proper title, give his or her initials (or forename, if you wish) and indicate his or her organization or department, if appropriate. The statements can be written as a list or in paragraph format.
- An acknowledgement of spiritual assistance (in the form of a tribute to God or Allah, for example) is best conveyed in worship and not written in the report.
- The acknowledgements are usually located on their own page after the **abstract** in a thesis; however, in a technical report or an academic or journal paper, the acknowledgements are usually placed after the main body of the report or paper (i.e., after the **conclusions** or **recommendations** but before the **references**). This part of the report is not considered as a **chapter** and would not be numbered.
- See also **Dedication**.

Acronyms

See **Abbreviated terms**.

Aims

See **Objectives**.

Alphabetical arrangement (of lists)

- *Groups of words*: Two systems are used to arrange groups of words in alphabetical order:

    (1) *Word-by-word*: The arrangement is alphabetical by the first word of each entry. As a space comes before a letter, *Old Sarum* precedes *Oldcastle*. Entries that begin with the same word are arranged in alphabetical order by the second word (and so forth), thus *New England* precedes *New Zealand*. As nothing comes before something, *Clare* precedes *Clarecastle*.

    (2) *Letter-by-letter*: The arrangement is alphabetical with every letter considered up to a "delimiting" punctuation mark (i.e., period, comma, colon, semicolon, or parenthesis), but ignoring spaces; thus *Eastleigh* precedes *East London*.

    In both systems, punctuation marks such as hyphens, dashes, slashes, quotation marks, or apostrophes are ignored; thus *L'Aquila* precedes *Lausanne* and *Carrickmacross* precedes *Carrick-on-Shannon*.

- *Abbreviations and symbols*: An alphanumeric character-by-character sorting may be used for lists of abbreviations, acronyms, mathematical and chemical symbols, and so forth, with all entries included in a single list (e.g., C, C_2H_4, C_4H_{10}, $CaCO_3$, C/B, CNC, CO_2) or separated into categories.

- *Personal names*: The arrangement is typically alphabetical by surname; entries with the same surname are alphabetical by forename(s). Names that have different forms, for example: Mac—, Mc—, M^c—, MC—, are best alphabetized as given (and not as if the names were spelled out).

- *Articles in titles*: Articles (e.g., a, an, the) appearing at the start of the titles of published works, for example, may be ignored if they are not part of a proper name.
- *Computer sorting*: Word processors and spreadsheets (and other computer programs) sort using the ASCII (American Standard Code for Information Interchange) character set. Sorting is by a strict coding of characters—which is useful as it always gives the same answer—but it can result in apparent anomalies as punctuation marks, spaces, capitalization, special symbols, and characters (e.g., @, $, &, ®, %) are all included in the sorting process.

Annex

An annex comprises supplementary documents bound at the end of the report. It has a similar, but not identical, role to an **appendix**. Whereas an appendix contains supporting information directly linked to elements in the main part of the report, an annex typically comprises self-contained documents, which may be available elsewhere, but are included in the report for the convenience of the readers. For example, a consultant's report, submitted following the assessment of safety standards and procedures in a factory, may include, as an annex, the company's safety policy.

APA Manual

- The *Publication Manual of the American Psychological Association* (6th ed., American Psychological Association, 2010) is a highly regarded manual of editorial style, used by many individuals and organizations to present written material in the social and behavioral sciences. The APA's style for the citation of references is also used in other academic disciplines, including science and engineering.
- In regard to the format for citing references using the popular author-date method, a few stylistic differences exist between the convention

outlined in **Citing references (examples of author-date method)** and the APA's style. The main differences are:

(1) A comma is used to separate the name(s) of the author(s) and the year in a parenthetical citation (see examples below).

(2) An ampersand is used to link the names of authors when they appear in parentheses (i.e., in preference to the word *and*).

(3) When a work has three, four, or five authors, the names of all authors are given the first time the reference appears and thereafter, for subsequent citations of the same work, only the lead author is given followed by *et al.* However, for works of six or more authors, only the lead author is mentioned (followed by *et al.*) in the first and in all subsequent occurrences.

- Examples of APA-style reference citation (author-date method):
  - Single author mentioned in the text: *The metacentric height of the tub was estimated using the method of Butcher (2002).*
  - Single author not mentioned in the text: *The metacentric height of the tub was estimated (Butcher, 2002).*
  - Two authors not mentioned in the text: *The metacentric height of the tub was estimated (Butcher & Baker, 2003).*
  - Three or more authors not mentioned in the text: *The metacentric height of the tub was estimated (Butcher et al., 2004).*
  - If the page number (or another element of the reference work) were to be mentioned, this would follow the year: *The metacentric height of the tub was estimated (Butcher et al., 2004, p. 329).*
- See also **Style manuals** and **Citing references (basic rules)**.

Appendix

- An appendix (plural: appendices) contains relevant supporting information that is not regarded as essential to the comprehension of the main report. As many readers will not have the time (or the interest) to read the complete appendix, ensure that all important results are presented in the main report (with appropriate reference to the supporting details in the appendix).
- If the material that you wish to include can be separated into a number of themes or sections, it is best to create individual appendices.
- Each appendix should have a title and be identified by a capital letter or number (ideally allocated sequentially in the order in which they are mentioned in the report, but this is not critical). Capital letters are preferred as they enable a distinction to be made between chapters in the main report (which are allocated numbers) and the appendices.
- It is a good idea to start each appendix with a brief introduction, which can describe the content and place the material in context within the report.
- The appendices are usually paginated separately from the main text and from each other. It is common practice to number the pages in Appendix A as A1, A2, A3, etc. This is a big help in compiling a large report, as the individual appendices can be finalized and put aside before the main report is completed. Remember to number all pages, including material extracted from other sources.
- Appendices containing extensive and detailed mathematical calculations can be written by hand; typing equations is time-consuming and there is a strong temptation to skip out steps (making it difficult, or even impossible, for readers to understand what was done). If the report is to be stored electronically, you could scan the written material. This approach would be acceptable for most academic or industrial reports, but for a thesis, there may be no option but to type the material (do not leave out crucial steps, though).
- See also **Annex**.

What goes into an appendix?

- Never use the appendix as a dumping ground for bits of unwanted text that you could not find space for in the main report. It is also a bad idea to throw in a technical standard, for example, just to expand the thickness of the report (size is not a measure of writing quality).
- A pragmatic rule: do not place items in the appendix that are readily available in the public domain (e.g., standards established by well-known organizations, journal papers, extracts from textbooks), but you could, if you wish, include non-archival documents (such as equipment specifications or company product sheets, but be careful not to infringe copyrights).
- In addition, the appendix could include
  - A design specification (supplied by the client, for example) that was the baseline for the study;
  - Tables of raw data;
  - Extensive mathematical derivations or statistical analysis;
  - Examples of calculations that yielded results presented in the main part of the report (usually it is sufficient to include typical or illustrative examples, rather than all of the workings);
  - A description of novel equipment or apparatus manufactured for the study;
  - A glossary; or
  - A list of symbols and abbreviations and their definitions (such information is, however, more commonly placed after the table of contents).

ASCII

- ASCII stands for "American Standard Code for Information Interchange." The standard ASCII character set consists of 128 numbers, which are assigned to letters, numerals, punctuation marks, and the most commonly used special characters. The Extended ASCII character set represents additional special characters (e.g., mathematical symbols) and foreign language characters.

- A document in "ASCII format" is in "plain text," that is, without any text formatting parameters such as line spacing, tabs, special fonts, or underlining. The format is convenient for moving blocks of text from one application to another (e.g., into and out of web documents). "ASCII format" files can be created using Microsoft Notepad or by saving the file as "Text Only" in Microsoft Word.
- Further details on ASCII can be found in ISO 14962:1997 (Space data and information transfer systems – ASCII encoded English).

Assessment report

An assessment report is a document that records findings and makes recommendations following the assessment of the capabilities or condition of an artifact or individual (e.g., the structural condition of an old stone bridge or the speech difficulties of a stroke patient). The typical structure of an assessment report follows.

Typical structure of an assessment report

Title page
Summary
1 Introduction
  1.1 Background
  1.2 Details (name, address, date of assessment/inspection, etc.)
2 Purpose *or* Reason for assessment/inspection/referral
3 Tests conducted *or* Assessment process *or* Assessment procedure
4 Results
5 Conclusions *or* Conclusions and recommendations
6 Recommendations
References
Appendix

Assumptions and approximations

- It is the treatment of gray issues such as assumptions and approximations that distinguishes good technical writing. Never gloss over these issues: they should be clearly described in the report and not buried deep in an appendix. Your report will be deemed a failure if someone, in attempting to replicate your results, is unable to do so because you failed to fully explain an assumption or to clearly describe an approximation that you made.
- It is important to consider the consequences of each assumption or approximation; this can be presented in the **discussion**. For example, the frequently used "small angle" approximation: $\cos \alpha \approx 1$, works exceptionally well when $\alpha = 5°$, producing an error of less than 0.4%; however, if $\alpha = 15°$, then the error grows to 3.4%. That may still be perfectly acceptable, but it is important to assess this for the particular study undertaken.
- See also **Bad results** and **Speculation**.

B

Background

The **introduction** frequently leads into two descriptive elements of the report: the **background** and the **literature review**. These elements provide a history of what has been done before. The background should place your work in context for the reader, describing relevant external constraints or events that were important for the study. It provides a link between what has been done before and what is described in your report.

Frequently an adequate description of the background and literature review can be completed in a few paragraphs and contained under the heading *introduction*. However, if the background information that must be provided—to enable the readers to comprehend the report—is extensive, then it should be written as a chapter in its own right and located immediately after the introduction. For a thesis describing original work, it is usually necessary to conduct a thorough literature review, and this too would typically be written as a separate chapter.

Bad results

Do not fall into the trap of thinking that only "good" results should be included in the report. Provided that the experimental work was soundly conducted, the results, even if they were a dismal failure in terms of what you set out to do, are worthy of reporting. By recording "poor" results you give other researchers the benefit of your experience. These results should lead into a discussion of probable causes of the "failures," and you can then make suggestions on how to get it right next time.

There is also a chance that what you think is incorrect—because it does not fit into conventional theory or match previously reported results—is actually correct. You may not have sufficient evidence to prove this, but it is nevertheless good practice to include such results in your report; just be honest about your understanding of the results (see **Honesty**).

Bibliography

- A bibliography is a list of books and other written material on a selected subject containing details on authors, titles, dates of publication, and so forth. It may also contain brief comments on the usefulness or content of the particular work. This is not the same thing as a **literature review**—often included as part of the introductory material in a research report or thesis—which reviews the state of the art, describing previous work, with relevant comments regarding, for example, contradictions, gaps, and limitations in the literature.

- The term *bibliography* is also used for a list of works consulted during a study, but be careful here: it does not have the same meaning as the term *references*, which should be used for a list of works cited in a report (see **Citing references (basic rules)**).

- The inclusion of a bibliography, instead of a *list of references*, is not popular in technical documents. It is occasionally included after the list of references (in which case only additional works, not mentioned in the references, would be listed); however, this information is frequently of no value to the readers (and if it serves no purpose, it would be better to omit it).

- There are many conventions or styles prescribing the sequence and format of the publication details—information that readers will need to find the works listed. The format described for the numeric method of citation—see **References (basic rules)** and **References (examples of numeric method)**—may be used for generating a bibliography.

- If you need more information on bibliographies and citing references using footnotes/endnotes (i.e., humanities style), consult the publications listed under **Style manuals**.

C

Capitalization

This is a literary minefield; use a good dictionary to navigate, or you risk being blown up. The details and examples given below are illustrative and are not intended to be comprehensive.

Capitalized names and terms (initial letter is a capital)

- Capitalize names of persons. Be careful with European names that have articles such as *de*, *du*, *von*, *van der*, and so forth, as they are mostly written lowercase (e.g., de Gaulle, du Prez, von Braun, van der Merwe), but not always. Check carefully to avoid offense or misrepresentation.
- Capitalize titles preceding personal names (e.g., President Cooper, Colonel Abu Zafar, Doctor Obasanjo, Professor Mueller, Sir Arnold, Mayor O'Shea), but not when the title follows the name, as a "job description" (e.g., Gina Piacentini, senator; Dickie Flynn, bishop of Newtown). When reference is made to a person using the title (e.g., the president, the doctor, the professor), then lowercase is used. However, in an address to such a person, when the title is used on its own, then it is capitalized (e.g., Dear Professor …).
- Capitalize names of places (e.g., cities, rivers, mountains, geographical regions, countries), structures, and public places (e.g., bridges, buildings, streets, parks), organizations (e.g., government departments and bodies, judicial bodies, political parties, economic forums, companies, and associations), months and days of the week (but not seasons), holidays, religious titles and terms, scientific names of animals and plants, names of airplanes, trains, boats, and so forth. There are many more.
- Capitalize the names of organizations (e.g., at the University of Limerick, in the US Navy); however, when reference is made to the organization without mentioning the name (e.g., at the university, in

the navy)—say, in a generic way, following the initial mention of the organization—lowercase is used.

- Capitalize nouns formed from proper names (e.g., Mach number but not Mach Number, Laplace transform, Coriolis force, Bernoulli's equation, Coulomb's law), with the following exceptions: units of measure (e.g., kelvin), chemical elements (e.g., rutherfordium), minerals (e.g., fosterite), and atomic particles (e.g., bosons).
- Capitalize the names of celestial bodies in the universe, such as planets (e.g., Jupiter—although Earth is frequently written lowercase), moons (e.g., Phoebe, a moon of Saturn—although the Earth's moon is usually written as *the Moon*), comets (e.g., Hale-Bopp), stars (e.g., Alpha Centauri), constellations (e.g., Ursa Major), and galaxies (e.g., Crab Nebula).

Capitalization in sentences (including lists)

- The first word of each independent sentence should be capitalized.
- A complete sentence in parentheses, when enclosed within another sentence, does not require an initial capital letter, unless a terminal punctuation mark (question mark or exclamation point) is used.
- Material introduced by a colon (such as an explanation or elaboration) comprising more than one complete sentence should be capitalized (see **Punctuation**, section 2.1).
- List entries (or items) contained within a sentence are usually not capitalized.

  Example:

  > The student was required to (1) paint the steel disk black, (2) place it 100 mm below a 60 W light bulb, and (3) measure the temperature change.

- Entries (or items) in vertical lists, whether bulleted or numbered, can start with either an upper- or a lowercase letter, but complete sentences should be capitalized (the adopted format should be consistently applied throughout the report).

Capitalized terms for parts of a report

When referring to a specific named figure (or table, chapter, section, appendix, equation, or theorem), it is acceptable to capitalize the initial letter (e.g., see Table 3.1 and Figure 3.4, refer to Appendices B and D). This is a matter of style preference, but capitals should not be used for minor report elements (such as a column number or a sample number) or when the words are used in a generic sense, without referring to a named part of the report (e.g., earlier chapters, tables of data).

Titles, captions, and headings

- Titles of published works, figure captions, table titles, and chapter and section headings are, by convention, written in one of two ways:
  - (1) *Title case*: the initial letter of the first word and the first letters of the main words are capital letters; main words in this context exclude conjunctions (e.g., and, but, for, if, or, so), articles (e.g., a, an, the) and prepositions (e.g., above, across, after, at, below, by, down, for, from, in, into, of, off, on, over, since, to, up, with); however, some **style manuals** recommend capitalizing prepositions of four or more letters.
  - (2) *Sentence case*: the initial letter of the first word is a capital letter, followed by all words in lowercase, except for acronyms, certain abbreviations and proper nouns.
- It is important to apply the adopted format consistently throughout the report.

Full caps (all uppercase)

Full caps are used for computer code and certain abbreviated terms (some acronyms, however, such as *laser* and *radar*, have become so familiar that they are now ordinary lowercase words). Full caps may also be used for headings of chapters, sections, columns in tables, et cetera. Note that the use of full caps for emphasis is dreadful style: it resembles SHOUTING and is more difficult to read.

CBE/CSE Manual

Scientific Style and Format: The CSE Manual for Authors, Editors, and Publishers (7th ed., Rockefeller University Press, 2006) is an acclaimed manual of style for the physical sciences produced by the Council of Science Editors, which was formerly known as the Council of Biology Editors (CBE). It provides recommendations for publishing journals, books, and other scientific material, covering both American and British English style preferences.

See also **Style manuals**.

Chapters and sections

The report is divided into chapters (i.e., major subdivisions), and each chapter may be further divided into sections and sub-sections (i.e., minor subdivisions). Each chapter would usually start on a new page. Chapters and sections should be numbered; see **Numbering system (for chapters and sections)**. Material that appears before the introduction (i.e., the *front matter* in the language of publishers, which includes title page, abstract, and table of contents) is not divided into chapters and should not be numbered.

See also **Headings** and **Structure of reports and theses**.

Characteristic numbers

A list of characteristic numbers commonly encountered in engineering is given in the table that follows. Each number is represented by a two-letter symbol. When these symbols appear as a product in an equation, it is helpful to separate the symbols from other symbols by a space, multiplication sign or parenthesis.

Characteristic numbers

Symbol	Name	Symbol	Name
Al	Alfvén number	*Nu*	Nusselt number
Co	Cowling number	*Pe*	Péclet number
Eu	Euler number	*Pr*	Prandtl number
Fo	Fourier number	*Ra*	Rayleigh number
Fr	Froude number	*Re*	Reynolds number
Gr	Grashof number	*Rm*	magnetic Reynolds number
Ha	Hartmann number	*Sc*	Schmidt number
Kn	Knudsen number	*Sr*	Strouhal number
Le	Lewis number	*St*	Stanton number
Ma	Mach number	*We*	Weber number

Notes:

1. See ISO 31-12:1992 for further details.
2. The symbols are usually set in italic typeface.
3. When the names are written out, roman (upright) typeface is used and the names are capitalized (as shown in the table).

Charts

- Charts (*x*-*y* graphs, bar, and pie charts, for example) are treated as **figures** in the report. All figures should be given a unique figure number and a caption, which are placed below the figure.
- Always acknowledge the source of the data when you were not responsible for its collection or production (failure to do so is **plagiarism**).
- Charts without (1) appropriate labels on the axes and data series, and (2) correct units, are practically useless—hence, you should never omit this information.
- Use grid lines and/or scales on *x*-*y* graphs to assist readers to interpret the data—you cannot always anticipate what your readers will find interesting, so it would be helpful if they could estimate the coordinates of the data points (without having to draw lines on the graph and scale off the values).

- Write the numerals on the axes in a consistent manner—for example, do not indicate a series as *0.5*, *1*, *1.5*, *2*; rather write *0.5*, *1.0*, *1.5*, *2.0*.

- For *x-y* graphs, an offset origin (i.e., $x \neq 0$ and/or $y \neq 0$) can mislead readers, particularly when the *x* and/or *y* value of the origin is close to, but not equal to, zero. This should be avoided, if possible.

- Do not join data points on an *x-y* graph if you have only two points, unless there exists theoretical or experimental information about the *x-y* relationship that can be used to support the drawing of a straight line or curve between the points.

- An extrapolation on an *x-y* graph (beyond the data points) should be indicated by a dashed line.

- Only continuous functions should have the data points joined on an *x-y* graph. Non-continuous data should be shown as discrete points, or alternatively—as is shown in the example below for annual rainfall—as a bar chart.

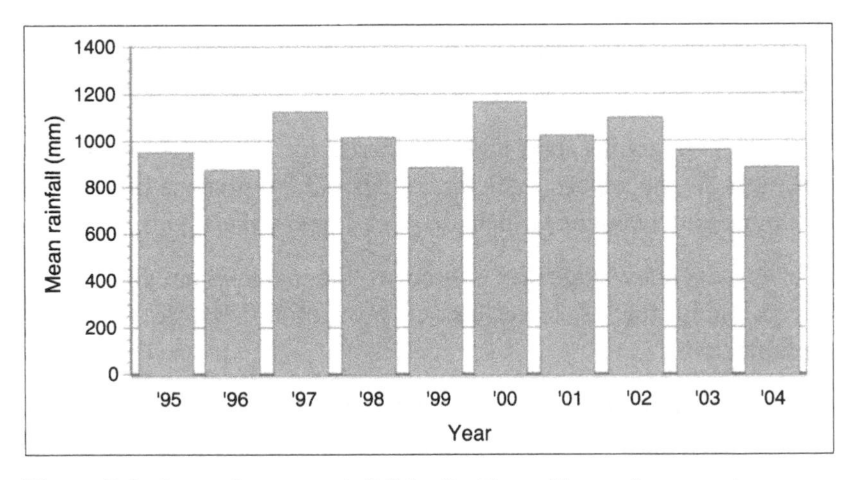

Figure 7.2 Annual mean rainfall in the Turnpike region over ten years (data from Turnpike Airport Authority 2005)

- Use the same units, symbols (icons), and colors for all charts in the report that display the same or similar kinds of data. Similarly, it is desirable to have the same scales and grids on charts that are likely to be compared to each other (although this may not always be practical).

- If it is likely that the report will be reproduced using a black and white copier, use unique symbols or shading to identify the data series rather than color.

Chemical elements, compounds, and symbols

- A list of chemical elements and their symbols is given in the table that follows. The names of the elements are not capitalized—and this includes elements named after people or places (e.g., californium). The symbols, however, have initial capital letters and are written in roman typeface without periods (e.g., Cf). Chemical compounds, when spelled out, are written in lowercase without hyphens (e.g., aluminum oxide) and element suffixes are set as subscripts (e.g., Al_2O_3).

- The full name of the chemical element or compound is normally used the first time it is mentioned in the report, followed by the chemical symbol or its abbreviation in parentheses—examples are: *nitric acid (HNO_3)* and *tetraethylammonium-p-toluenesulfonate (TEATS)*. The symbol or acronym can be used thereafter, if you wish.

- The symbols should be used appropriately in passages of text; in many cases it is better to write out the chemical name—for example, say: *Atmospheric water and ozone levels were measured*, rather than *Atmospheric H_2O and O_3 levels were measured.* Do not use element symbols as "shorthand" in the text (e.g., Au fish).

- Concentrations of solute to solvent, when expressed as a percentage, should always indicate if it is a volume–volume ratio (i.e., vol/vol) or a weight–weight ratio (i.e., wt/wt)—for example: *80% (vol/vol) ethylene glycol* or *80% v/v ethylene glycol.*

Chemical elements and their symbols

Symbol	Name
Ac	actinium
Ag	silver
Al	aluminum, aluminium[2]
Am	americium
Ar	argon
As	arsenic
At	astatine
Au	gold
B	boron
Ba	barium
Be	beryllium
Bh	bohrium
Bi	bismuth
Bk	berkelium
Br	bromine
C	carbon
Ca	calcium
Cd	cadmium
Ce	cerium
Cf	californium
Cl	chlorine
Cm	curium
Co	cobalt
Cr	chromium
Cs	cesium, caesium[2]
Cu	copper
Db	dubnium
Ds	darmstadtium
Dy	dysprosium
Er	erbium
Es	einsteinium
Eu	europium
F	fluorine
Fe	iron
Fm	fermium
Fr	francium
Ga	gallium
Gd	gadolinium
Ge	germanium
H	hydrogen
He	helium
Hf	hafnium
Hg	mercury
Ho	holmium
Hs	hassium
I	iodine
In	indium
Ir	iridium
K	potassium
Kr	krypton
La	lanthanum
Li	lithium
Lr	lawrencium
Lu	lutetium
Md	mendelevium
Mg	magnesium
Mn	manganese
Mo	molybdenum
Mt	meitnerium
N	nitrogen
Na	sodium
Nb	niobium
Nd	neodymium
Ne	neon
Ni	nickel
No	nobelium
Np	neptunium
O	oxygen
Os	osmium
P	phosphorus
Pa	protactinium
Pb	lead
Pd	palladium
Pm	promethium
Po	polonium
Pr	praseodymium
Pt	platinum
Pu	plutonium

Chemical elements and their symbols (*continued*)

Symbol	Name	Symbol	Name
Ra	radium	Ta	tantalum
Rb	rubidium	Tb	terbium
Re	rhenium	Tc	technetium
Rf	rutherfordium	Te	tellurium
Rh	rhodium	Th	thorium
Rn	radon	Ti	titanium
Ru	ruthenium	Tl	thallium
S	sulfur[3]	Tm	thulium
Sb	antimony	U	uranium
Sc	scandium	V	vanadium
Se	selenium	W	tungsten
Sg	seaborgium	Xe	xenon
Si	silicon	Y	yttrium
Sm	samarium	Yb	ytterbium
Sn	tin	Zn	zinc
Sr	strontium	Zr	zirconium

Notes:

1 Source: Periodic Table of the Elements, International Union of Pure and Applied Chemistry (IUPAC), Research Triangle Park, NC, USA. www.iupac.org/reports/periodic_table/. Web page modified: Jan. 19, 2004, website accessed: Feb. 26, 2004.

2 British English spelling.

3 The British English spelling *sulphur*, widely used in non-scientific writing, is recommended neither by the IUPAC nor by the ISO (International Organization for Standardization).

Chicago Manual of Style

The *Chicago Manual of Style: The Essential Guide for Writers, Editors, and Publishers* (15th ed., University of Chicago Press, 2003) is an authoritative tome that may be used to end arguments on issues of general writing style, punctuation, grammar, citation of references, and so on. It reflects American English style preferences and is used worldwide for general publishing (see **Style manuals**).

Citing references (basic rules)

- This is a subject prone to nit-picking by academics (arguments concerning the relative merits of different citation styles are passionately voiced). To be honest, it is a rather boring topic; nevertheless, it is one that you must take seriously if you want to produce a quality piece of work.
- Citing references in the text enables the readers to distinguish between what you, the author, did—or dreamed up—and what you read. It provides the link between statements, results, figures, and so on, that you wish to credit to someone else, and the publication details (i.e., author, title, publisher, etc.), which are usually placed in a list at the back of the report. It enables the readers to identify, and obtain if they wish, the work to which you have made reference.
- There are a number of styles that get used for citing references; they fall into three categories:

  (1) Author-date method (with details of cited references given in an alphabetically ordered reference list).

  (2) Numeric method (with details of cited references given in a numerically sequenced reference list).

  (3) Footnotes/endnotes (with or without details of cited references given in a bibliography).

Citing references: author-date method

- The author-date method (which is also known as the parenthetical or Harvard method) uses the author's name followed by the year that the work was published—for example: *Polygraph results are considered inadmissible as legal evidence (Gepetto 2002).* A comma is often used to separate the name and the year.
- The author's name (i.e., family name only, without title) and the year are put in parentheses and placed before the period. Alternatively, if the author is mentioned in the sentence, then the year is simply placed, in parentheses, after his or her name—for example: *The*

technique of magnetic-resonance imaging of the brain was tested by Cricket (2003) to probe subjects' honesty.

- More examples are given in **Citing references (examples of author-date method)**. Details of cited references are given in a reference list (alphabetically ordered by author's family name) and not as footnotes/endnotes—see **References (basic rules)**. If you use the author-date method do not allocate numbers to the references.

Citing references: numeric method

- Each reference is given a unique number. These can be allocated either consecutively—that is, in the order that they appear in the text (called citation-sequence)—or alphabetically by the family name of the author (in which case the numbers will not be sequential in the text). The number is written in brackets or parentheses.
- The number can be placed at the end of the sentence or at an appropriate location in the sentence, such as after the author's name—for example: *Dumpty [14] investigated the crashworthiness of a semi-monocoque structure as an improvement to the monocoque shell.*
- It is extremely popular to write the number as a superscript, with or without brackets or parentheses—for example: *It was reported by Kingsman*[(15)] *that the monocoque shell was damaged beyond repair.*
- More examples are given in **Citing references (examples of numeric method)**. Details of cited references are given in the reference list and not as footnotes/endnotes—see **References (basic rules)**.

Citing references: footnotes/endnotes

- Superscript numerals—starting with 1 for the first note and numbered consecutively—are used to cite references; for example: *The results of field studies conducted by De Locks*[1] *into the dietary habits of Ursus arctos are considered.*

- Reference details of the cited work are given in either a footnote (i.e., on the same page) or an endnote (i.e., at the end of the section or chapter), and full details are often given in a **bibliography**. All footnotes/endnotes are numbered in sequence, resulting in citation notes being interspersed with explanatory text notes.

- If a work is cited more than once, the subsequent notes may repeat the reference details or, as is more popular, refer the reader to the earlier note (see **Ibid.** and **Op. cit.**).

- The page number (or section, chapter, table, or figure number) of the appropriate reference may also be indicated in the note—for example, if both note 1 and note 12 refer to pages in the renowned work of De Locks, then note 12 could state: *12. De Locks, ref. 1, p. 11.*

- This method of citing references, traditionally favored in the humanities, is seldom used in technical or scientific writing. It has many subtle rules, as described in, for example, the *MLA Style Manual and Guide to Scholarly Publishing* or the *Chicago Manual of Style* (see **Style manuals**).

Citing references: selecting a method

- Use the citation method favored by your supervisor (or organization). In the science and technology disciplines—and increasingly in many social sciences—the choice is between the author-date and the numeric method.

- The *author-date* method is the simplest to implement. It also saves the readers' time, as they do not have to repeatedly turn to the reference list to find the names of cited authors (it is a good idea to make life easy for your readers, who may be familiar with the references used).

- The *numeric* method, however, is more compact and is a better option if you wish to quote multiple references at the same point in the report. Journal editors are fussy and space is at a premium: their preference is almost always the numeric method.

- Whichever method you choose, stick to it rigidly and do not switch between one method and another. The great thing about getting this right is that the report will look professional even if the content is mediocre.

Citing references (examples of author-date method)

The following are examples of author-date citations as they would appear in the text. Full details (i.e., author(s), title, publisher, etc.) would be given in the reference list. See **References (examples of author-date method)**.

Citing references: one publication

- *Randomized controlled trials conducted by Sniffel (2004) supported the notion that happy children get fewer head colds.* This is correct for a single-author work, when the author is mentioned in the sentence.
- *Beneficial effects of omega-3 fatty acids—most easily obtained by eating fish—on cardiovascular disease (CVD) have been reported (Herengus and Pollock 2003).* The single reference has co-authors (i.e., Herengus and Pollock), which are not mentioned in the sentence.
- *Somnambulism, if infrequent, is not considered a sign of mental disorder (Grumpy et al. 2002).* The abbreviation *et al.* means *and others*. It is used when there are more than two authors (three in certain referencing styles, for example: Chicago Manual, MLA, Turabian). The names of all authors are given in the appropriate entry in the reference list.

Citing references: two or more publications

- *High pumping efficiencies have been demonstrated using an Archimedes screw design (Jack 2000; Gill 2001).* Multiple references are separated by a comma or semicolon.

- *The mathematicians Rex and Collie (1989; 1992; 1993) jointly developed numerical models for predicting cyclonic weather systems.* A comma or semicolon is used when an author (or co-authors, as in this example) has multiple works cited.

Multiple references of the same author(s) in the same year

- When two or more references have the same author(s) and year, they are distinguished by lowercase letters (a, b, c, etc.) added after the year of publication, allocated in their order of appearance in the report—for example: *The method of Phoolproof (2001a) was used to determine the loads on the spacecraft.* The letters also appear after the year in the corresponding entries in the reference list.
- When two or more works, published in the same year, have the same lead author, but different collaborating authors, the same technique of appending a lowercase letter to the year may be used. For example, a work by Klein, Petit, and Grande (2003); and one by Klein, Pequeno, Gross, and Petit (2003) can be cited as *Klein et al. (2003a)* and *Klein et al. (2003b)* respectively—with the corresponding details given in the reference list. Alternatively, and this is favored in certain **style manuals**, the second author (and third if needed) is named in the citation to distinguish the two works.

Different authors with the same name

- To distinguish between works published in the same year by authors with the same family name, the author's initials should be given—for example: *The effectiveness of the aural lure surpassed expectation (P. Piper 1987).*

Reference with page, figure, table, section, or chapter number

- If you wish to refer to a specific part of a publication (as would be the case when data has been extracted from a handbook), indicate this after the year—for example:

- *According to the Daedalus Handbook of Hydrocarbon Waxes (2003, p. 896) the strength of alkane wax reduces significantly when the temperature exceeds 105 °F.*
- *The surface temperature of the wing reached 150 °F (Icarus 2004, Table 5.2).*

Reference with no obvious author

- In this context the "author" is the name under which the work is listed in the reference list. It can be the editor or the name of the organization or company responsible for the work—for example: *The eating of shellfish has been shown to lead to increased juvenile attentiveness (Department of the Marine 2004).*
- It is not necessary to give the full name of the organization as it will be given in the reference list; if an acceptable abbreviation exists for the organization (e.g., IAEA, UNESCO), use it to reduce space. If no details are available concerning the author, editor, or organization, then as a last resort use an abridged title of the work. However, if the author of the work is designated as "Anonymous," then it is acceptable to use the word *Anonymous* (or the abbreviation *Anon.*) as the name of the author in both the citation and the reference list.

Reference with no date

- Use the abbreviation n.d. (no date) in place of the year—for example: *Early detection of amnesia in Ovis aries has led to increased farm yields (Bowpeep n.d.).*
- If a work is in the process of publication (e.g., a journal paper that has been accepted by the editor, but is not yet in print), write the words *in press* in place of the date in the citation and in the corresponding entry in the reference list. However, a work that is still under review (pending a decision on publication) should not be treated as *in press*, but rather as an unpublished manuscript; in this case the date on the manuscript should be used or, in the absence of a date, the abbreviation *n.d.* should be used. Such unpublished works are re-

garded as non-archival—see **References (non-archival)**—and it is author's details (not the prospective publisher's) that should be given in the reference list.

Reference in parenthetical material

- It is not necessary to put the year in parentheses or brackets when an author (or authors) is mentioned in parenthetical material—for example: *Amnesia in Ovis aries is regarded as rare (see Buoyblu 2000 for a comprehensive assessment of this troublesome condition).*

Citing references (examples of numeric method)

- The reference numbers are either allocated consecutively (i.e., in the order that they appear in the text) or according to the alphabetical sequence in which they appear in the reference list (this is less popular).
- The reference numbers may be written in brackets or parentheses in line with the text or as a superscript—again in brackets or parentheses or on their own. Choose one style and use it consistently throughout the report. The number may be placed at the end of the sentence or at an appropriate point in the text—for example, following the name of an author or after a statement or fact that you wish to reference.
- The following are examples of numeric method citations—illustrating the various options—as they would appear in the text. Full details (i.e., author(s), title, publisher, date, etc.) would be given in the reference list. See **References (examples of numeric method)**.

Citing references: one publication

- *Farmer*(14) *contends that severing a mouse's tail can lead to disorientation and visual disturbance.*

- *Milking cows had a negligible impact on the greenhouse gas output of the state [2].*
- *The experimental work of Mamba*[32] *demonstrated that a high fat diet and little exercise results in weight gain in caged rodents.*
- *Anabolic steroids, which were banned in 1974 by the IOC for use by athletes*[[21]]*, were identified in pre-prepared gingerbread.*

Citing references: two or more publications

- *The studies conducted by Laural [17] and Hardey [18] were inconclusive.*
- *The preference for marsupial dung by native dung beetles caused problems for Australian cattle farmers*[3,4]*.*
- *Recent publications*[[8,10–12]] *on predicting the height of the dead cat bounce failed to account for substrate stiffness.* The reference numbers are usually written "closed up," without spaces.

Reference with page, section, or chapter number

- If you wish to identify a specific part of a publication (e.g., page, section, chapter, or figure number), do so after the reference number —for example: *The Malachite Kingfisher is solitary (23, p. 463).*
- This is good practice when referring to encyclopedias, handbooks, compilations of data sheets, standard reference works, and so forth, but is not necessary for short documents (e.g., journal papers).

Citing references located on the Internet

See **Internet reference citation**.

Citing something second-hand

- This is a common dilemma: it arises when the author of the publication that you are reading has included something from another work (e.g., a figure, result, quotation) that you find useful and wish to incorporate into your report.
- There are three things to remember in this situation:

  (1) In almost all circumstances it is better to read the original work. Only cite something second-hand if you are unable to obtain the original work or if it is a translation and you are unable to read the original.

  (2) Always refer to the work that you read. Make it clear that what you are describing is indirectly based on the original work and acknowledge the author(s) of the work that you read—for example: *Hermes (2002) reviewed the experimental results of Sisyphus (1953) and noted that the "rolling coefficient of granite was approximately 0.072."* Alternatively: *The rolling coefficient of granite is approximately 0.072 (Sisyphus 1953, cited by Hermes 2002).*

  (3) Include details of both publications in the reference list. After giving the details of the original publication, refer to the work that you read—for example, by writing: *cited by Hermes (2002).*

- This approach ensures that the reader is able to identify the work that you read (it would also put the blame on the second author, that is, Hermes in the above example, if he or she messed up).

Citing standards

- Technical standards are treated a little differently than other references: give the full name of the standard in the text when the standard is first mentioned—for example: *The structural tests were performed in compliance with STN 2243-2 (Test methods for dental adhesives, Part 2: peel)*. Thereafter, you may refer to the standard by its numerical designation (i.e., STN 2243-2).

- For completeness, authors will often record full details of the standard in the reference list, including the name of the organization (and shortened address if it is not a well-known organization) responsible for the standard.

Concluding remarks

This chapter heading is sometimes used instead of **conclusions**. It allows greater flexibility to make statements that are less rigorously argued than would be the case for a conclusion. Conclusions should always be substantiated statements deduced from your results, whereas the heading *concluding remarks* allows some room for personal opinion.

Conclusions

- The concluding chapter of a report may have one of the following headings, depending on the intentions of the author:
  - Conclusions
  - Conclusions and recommendations
  - Concluding remarks
  - Summary of results

  The most widely used heading is *conclusions*—this is described below. The structure and content of **concluding remarks** and **summary of results** are described in their respective entries.

- The *conclusions* chapter should be a short, focused part of the report that highlights the essential results, conclusions, and deductions. The conclusions should cover all aspects of your work—not just the results, but also the methodologies (e.g., you could indicate how well the adopted approach worked or what the most critical or difficult aspects were).

- Note that there is an obvious repetition in the structure of the report, as all important points will already have been mentioned in the **discussion**. This is, however, not a problem as the purpose of the conclusions is different from that of the discussion, and the style of presentation is also different.

- A concise list of conclusions may be presented as a bulleted or numbered list. As whole pages of bulleted points may be difficult to read, paragraphs with sub-headings (or with the first words of each paragraph identified in bold or italic typeface) may be better in such cases. One advantage of numbering conclusions is that readers will be able to refer to specific entries with ease.

- Confine the conclusions to statements that are central to your work. Never conclude anything that has not been presented earlier in the report—that is, do not introduce new material in this chapter. In a list of independent conclusions, the main points should be stated first followed by the secondary points; in other words, get the most important information across first.

- There should be a synergy between the *objectives* and the *conclusions*: each objective should be matched to one or more conclusions, informing the reader whether the objective was met—and if appropriate, how it was achieved and to what extent.

- Note that a reader, after being enticed by the *abstract*, may then read the *objectives* and skip directly to the *conclusions*. The concluding statements should therefore stand on their own. References to figures, tables, and other results reported in earlier chapters may be helpful, but the concluding statements should be understood without the reader having to look elsewhere in the report.

- A common mistake—made by novices and experienced writers—is to overgeneralize conclusions. The constraints, assumptions, approximations and boundary conditions under which the study took place inevitably limit the applicability of the conclusions. Hence, conclusions should always be written in the context of the study: caveats, limitations, and restrictions should be clearly stated.

- It pays dividends to spend time on this chapter and to do it properly. You can get away with a badly worded sentence or an ambiguous statement elsewhere in the report, but try that in the conclusions and you will be heavily criticized. Review every sentence, with one question in mind: *Is this a conclusion?* It is surprisingly easy to end up with a *discussion* in the concluding chapter. Be ruthless and cut out every sentence that does not conclude something.
- See also **Logical conclusions**.

Confusables

See **Incorrect word usage (confusables)**.

Contents

See **Table of contents**.

Continuity

- Technical and scientific reports can be exceptionally difficult for readers not familiar with the specific subject area to understand. But, in many cases, it is not the technical jargon that poses the biggest problem (as the unfamiliar words can be looked up in a technical dictionary); it is the lack of adequate explanation.
- The problem frequently occurs when the writer shifts to a related topic after a paragraph break and fails to explain the context of the new topic. The lack of continuity may not be evident to the writer—or to a subject expert—but to a perplexed reader unfamiliar with the specific details, it can be baffling.
- A related problem involves the use pronouns (e.g., this, that, it, he, she, they). The pronoun would refer to something or someone in the preceding sentences (that something or someone is called the

antecedent); however, if this is not clear to the reader, he or she will be left asking: *What (or who) is the writer referring to?*

- Continuity is also important between sections in a report. A sudden change of topic from one section to another, or between chapters, can appear abrupt. A single sentence linking the topics, providing continuity and a smooth transition, can overcome this. Twists and turns are great in a novel, but readers of technical and scientific reports like to know where the writer is leading them.
- See also **Paragraphs**.

Conversion factors

See **Units (conversion factors)** for converting numerical values to, and from, SI units.

Copyright

Reproducing copyright material (defined below) in your report without permission—even if you have referenced it correctly—may be unlawful. The purpose of copyright is to give an exclusive right for certain uses of the work (see below) to the owner. There is, however, an important provision under the so-called fair-use (fair dealing) doctrine that permits a limited amount of copyright material to be reproduced or transcribed. Providing that the limits have not been exceeded (see **Fair-use doctrine**), permission from the copyright holder is not needed; nevertheless, to avoid charges of **plagiarism**, the source must always be acknowledged.

Copyright material

Copyright is a property right that applies to

(1) Original literary works (e.g., books, journals, web pages, computer programs, instruction manuals) and dramatic or musical works that have been recorded in writing or otherwise;

(2) Artistic works, including (a) graphic works (e.g., drawings, diagrams, maps, charts, plans, paintings, etchings, or similar works), and photographs (including images produced by radiation or other means), (b) works of architecture (buildings or models for buildings), or (c) works of artistic craftsmanship;

(3) Sound recordings or films (on any medium), broadcasts (including those on cable or satellite); or

(4) Typographical arrangements of published editions (comprising the whole or part of one or more literary, dramatic or musical works).

Copyright does not apply to

(1) Names, titles, or phrases (although these may be registered as trade marks);

(2) Ideas (copyright protects the way the idea is expressed, but not the idea itself); or

(3) Products or processes (e.g., industrial, theoretical, or experimental), although these may be protected by Intellectual Property (IP) agreements or patents.

Copyright: owner's rights

The owner of the copyright has the exclusive right to

(1) Copy the work (which means reproducing the work in any material form, including electronic means);

(2) Issue copies of the work to the public;

(3) Perform, show, or play the work in public, or broadcast it; or

(4) Make an adaptation of the work (which, for example, includes converting a computer program into or out of computer code or into a different programming language) or to do any of the above in relation to an adaptation of the original work.

The owner can also license others to exercise some or all of these rights.

Copyright life

A copyright has a limited life—in the US and Europe, for example, it expires 70 years after the death of the author or creator of a literary, dramatic, musical, or artistic work; for a sound recording, it lasts 50 years after it was made or released; and for a typographical arrangement of publications, it lasts 25 years after the edition was first published.

Copyrighting your work

Copyright registration is not required for copyright to exist (unlike trademarks and patents): copyright is automatic and effective from the time that the work is "fixed" in some way (that is, preserved or recorded on paper, CD, video, etc.; or placed on the Internet). It is a good idea to mark your work with the international copyright symbol © followed by your name and the year of creation. You may also wish to register your work with the US Copyright Office (see www.copyright.gov). The action is likely to deter potential infringers and may aid your case, if at some later stage you need to defend your copyright.

You should also be aware that the copyright of anything produced in the course of employment or under contract is likely to belong to the employer or contractor, unless an agreement to the contrary exists.

Copyright: public domain material

Material contained in works that are in the public domain may be used freely without permission. This includes works where the copyright has expired or where it failed to meet the requirements for copyright protection. It also includes most, but not all, US Government documents. Documents that were prepared entirely by a government agency are in the public domain; however, this may not be the case for documents that contain either (1) material resulting from work performed by a non-governmental organization or (2) copyright material (reproduced from another source, for example).

See also **Fair-use doctrine** and **Plagiarism**.

Criticism of others

- Treat others with respect—those with whom you disagree may have written something that is silly, but that does not necessarily mean that they are stupid, nor does it justify the use of derogatory terms. Present your data and your arguments, and let your readers come to their own conclusions regarding the stupidity of others.
- Always avoid personal criticism when you identify an error in the work of another researcher—for example, say: *The results obtained by the author do not agree with those of Pinocchio*[(13)], rather than *Pinocchio*[(13)] *made a serious blunder as is evident by the results obtained, and his conclusions are utter nonsense.*

Cross-referencing

- Good cross-referencing in the report makes it possible for readers to find their way around with the least effort and to absorb the critical elements with the lowest level of frustration.
- Get to the point promptly—for example: do not write *please refer to Figure 4.1 in the results chapter*, but *refer to Figure 4.1.* There is no need for *please*, and if the figures are properly numbered then the reference to the *results chapter* is redundant.
- Use parentheses—for example: *The maximum values of the seven sets of tests (Tables A.1 to A.7) are plotted in Figure 5.2 and correlated against the theoretical results (see Table 5.1).*
- Do not be lazy and omit details—for example: *The atmospheric readings presented earlier were used in the calculations.* Tell the readers exactly where the readings are recorded (i.e., the table number or page number) and indicate precisely which readings were used.
- Provide unambiguous cross-referencing statements—for example: write *see equations (2.4)–(2.6)* rather than *see equations (2.4–2.6).*

- Note that the words *figure*, *chapter*, *equation*, and *appendix* are usually spelled out, except when they appear in parentheses (and abbreviations are used)—for example: *Figure 6.3 revealed cracks in the material, which could not be seen on the corresponding micrograph (Fig. 6.4).*

CSE Manual

See **CBE/CSE Manual**.

D

Data presentation

- The exchange of data that is accurate and complete is a crucial element of engineering practice; well thought-out and concise **figures** and **tables** can make the difference between a mediocre and a good engineering report.
- Select the most important information for the report—data that you feel your readers would be most interested in seeing. (Deciding what to *leave out* can be an exceedingly difficult task.)
- There is no need to present the data twice (i.e., in a table and a chart). If the precise numbers are less important that the trends, just use charts.
- In the case of a bar chart, it may be possible to write the numbers above the individual bars, and in this way both the numerical and graphical information may be presented.
- In the text draw the readers' attention to each figure and table, but there is no need to dwell on a feature that is blatantly obvious just because you feel that you have to write something about it (see **Cross-referencing**).
- Be consistent throughout the report in the use of data symbols, colors, and units to represent the same or similar parameters in charts, drawings, and other graphical information.
- See also **Charts**, **Figures**, **Tables**, and **Statistical analysis**.

Dedication

Oh yes, this is the sloppy part—an extreme example: *I would like to dedicate this thesis to my beloved Patrick for his unquestioned devotion and love over the past twelve months.* (This over-the-top statement could be rather embarrassing if the author falls out with Patrick soon afterwards!) Typically the dedication is brief and simple (e.g., *To Fiona*).

A dedication is often included in academic works (such as a PhD thesis), but, in most cases, it would be inappropriate to include one in a workplace report. The dedication would normally be located on its own page after the title page (traditionally positioned centrally), and the heading *dedication* would not appear in the **table of contents**.

See also **Acknowledgements**.

Diagrams

See **Figures** and also **Hand-drawn sketches**.

Dictionary

Invest in a good one: do not rely solely on the word processor. See also **Technical dictionaries**.

Discussion

- This chapter is a *discussion*, so do not make it sound like a eulogy: engage the reader and write in a creative, natural way. The style and format of the discussion is less prescribed than that used in other chapters of the report, and you have plenty of freedom to discuss your interpretation of the **results** and their relationship to previous work. Always write in a positive way about your results, even if they are perceived to be "bad" (see **Bad results**).
- In the discussion you could explain
    - Why the results turned out as they did;
    - Why there were variances from your expectations;
    - Why your results differed from previously published work;

- How the experimental results correlated, or failed to correlate, with the theory;
- The consequences of approximations and assumptions; or
- The limitations of the study.

- It is important to choose your words carefully—for example: it is rare that a single study will conclusively *prove* something new; it is more likely that the results will *support* or *contradict* a theory (or hypothesis). Select appropriate words to indicate the strength of your assertion—for example: test results could *prove*, *demonstrate*, *show*, *indicate*, *support*, *suggest*, or *imply* something. Similarly, a recommendation could be *definitely*, *probably*, *possibly*, or *arguably* advocated, depending on your conviction of its merit.
- A framework that may be used to structure the *discussion* follows: you would typically
    - Introduce the chapter by briefly restating the main objectives;
    - Discuss each major element or aspect of the work (starting with the most important points), highlighting, for example, trends, limitations, contradictions, and so forth in the results; and finally
    - Present ideas and recommendations for future work (if appropriate).

E

Emphasis by typeface change

It can be incredibly irritating for a reader to encounter bold or italic words in almost every sentence in a report—the result of the writer's misguided attempt to create emphasis. The occasional use of a change in typeface is acceptable, but its overuse is distracting. Similarly, important words (or passages of text) should never be underlined or written in capital letters. Emphasis should come from the correct selection of words. There are exceptions: (1) italics are used for mathematical variables and symbols, titles of published works, scientific names (genus and species), foreign language words, and the names of individual boats, trains, aircraft, and so forth; and (2) underlining is used to identify URLs.

See **Capitalization** and **Italicization** and also **Style of writing [emphasis]**.

Endnotes

See **Footnotes and endnotes**.

Engineering report

- An engineering report provides a written record of a study that has been completed and contains a lot of detailed information and calculations. Such a record must be complete and unambiguous, permitting a peer to check and validate every single pertinent calculation. This point is extremely important: if the calculations cannot be validated the work is likely to be worthless.

- Structure the report to enable another engineer to check the work; this means that the sequence of chapters (and sections within chapters) must be logical, enabling the reader to follow—step by step—the work performed. It is vital that numerical calculations are presented accurately. Pay special attention to the presentation of spreadsheets: if the checker cannot replicate the calculations in your spreadsheets, the results will be of little value.

- Sometimes it is necessary, as part of a study, to read values from a chart contained in another reference, which may not be available to the reader. In this situation, it would be helpful if you reproduced the chart in your report—preferably in the **appendix**—and annotated it to indicate how you obtained the values that you have used; but be careful of infringing copyrights.
- Engineering reports may be prepared by one division within an organization for submission to another division, or simply to record the analysis completed before moving on the next phase of a project. They may also be prepared for submission to a government regulatory authority (e.g., the Federal Aviation Administration). Engineering reports must be accurate, complete and unambiguous, as the author may, in the event of an inquiry following a failure or accident, be legally accountable for the results presented.
- The structure of a typical engineering report is illustrated in **Table of contents**.

Epigraph

An epigraph is a short quotation—usually a poem or saying—included at the start of a literary work, or at the start of a chapter, intended to suggest a theme for what follows. It is usually written without quotation marks in italics and the author is credited below the quote. Epigraphs are popular in books and magazine articles, but are usually not recommended for technical or scientific reports or theses. If you wish to include one, be original in your selection.

Equations

See **Mathematical notation and equations**.

Errata

This is a list of errors in a report and their corrections. If errors are noted after the report is distributed, prepare a loose sheet with the title *Errata* for distribution to recipients of the report. Indicate the location of the errors (i.e., page and line numbers or equation numbers), the incorrect text or data, and the corrected versions.

Error analysis (measurements)

- When recording measurements, remember that all measuring devices have inherent inaccuracies. The errors are caused by system non-linearities, hysteresis, drift, resolution limitations, prolonged settling time, and so forth; there are also external influences such as environmental conditions (e.g., temperature) and operator bias—all of these factors can impact the precision, accuracy, and repeatability of measurements. The errors may be *random* or *systematic*; measurements are said to be *precise* if the random errors are small and *accurate* if the systematic errors are small. The total error of a measurement is the sum of the random errors and the systematic errors.

- *Random* (precision) errors produce measurements that are scattered around the true value (i.e., the theoretical mean of an infinite number of unbiased measurements). They are caused by random, or chance, events and are thus treated statistically. Of interest is the *standard deviation of the sample mean* ($S_{\overline{X}}$). This is a function of the spread of the individual measurements (i.e., the sample standard deviation, S_X) and the number of measurements taken (n). If x_i represents the individual measurements and $\bar{x}$ is the sample mean, then

$$S_{\overline{X}} = \frac{S_X}{\sqrt{n}} \qquad \text{where} \quad S_X = \sqrt{\frac{\sum_{i=1}^{n}(x_i - \bar{x})^2}{n-1}} \qquad \text{and } \bar{x} = \frac{1}{n}\sum_{i=1}^{n} x_i$$

Assuming a Gaussian distribution, it may be said, with a 95% confidence level, that the true value is within $\pm 2S_{\overline{X}}$ of $\bar{x}$. For small sample sizes (less than 30) the 95% confidence interval is wider—for example: if $n = 5$, the true value is within $\pm 2.78 S_{\bar{x}}$ of $\bar{x}$.

- *Systematic* (bias) errors produce measurements that have a systematic bias in a particular direction. They are caused by a variety of factors (e.g., a null error occurs when an instrument reading does not return to zero after a measurement). Such errors are cumulative.

- A popular method for combining errors (i.e., adding systematic errors or adding a random error to a systematic error) is the root-sum-square (RSS) method, which, for $A = f(x, y, z)$, is given by

  $\Delta A = \sqrt{\delta x^2 + \delta y^2 + \delta z^2}$ where δx, δy, and δz are the errors in x, y, and z respectively.

- The *probable error* of a function involving the sum (or difference) of n terms, that is $A = x_1 + x_2 + x_3 + ... + x_n$, is given by

  $\Delta A = \sqrt{\sum_{i=1}^{n} \delta x_i^2}$ where δx_i is the error in x_i.

  Similarly, the probable error of a function involving the product (or quotient) of n terms is given by

  $$\frac{\Delta A}{A} = \sqrt{\sum_{i=1}^{n} \left(\frac{\delta x_i}{x_i} \right)^2}$$

Et al.

- The Latin abbreviation *et al.* (from *et alii*) means *and others.* It is widely used in academic writing (and increasingly in other contexts). If used correctly it will give your report a professional touch, but be careful: the abbreviation is frequently written with an extra period (*et. al.*), and that is wrong.

- The abbreviation is used when citing a publication that has more than two authors (three authors in certain referencing styles, such as

Chicago Manual, MLA, and Turabian)—for example: *The number of insect splats on the windscreen was greatest in the months of June and July (Midge et al. 1996).* Having gained a few extra brownie points, do not throw them away by using poor grammar in a statement such as *Cicada et al. (1948) was the first to observe this.* There is obviously more than one author, so it should be *Cicada et al. (1948) were the first to observe this.* This is a common mistake.

- Unless there is a long list of co-authors (say eight or more), do not use the abbreviation in the reference list. (It may save you a few minutes by not having to record the names of all the co-authors, but think of the problem that you create for a reader if he or she has to source the work using an incomplete reference.)
- See **Citing references (basic rules)** and **References (basic rules)**.

Excerpts (extracts)

See **Quotations**.

Executive summary

- This heading is often used as a flashy alternative to *summary*. When used correctly, however, it is a concise overview of the report, written specifically for an audience who would not have the technical or scientific background to understand the report itself, or alternatively, are too busy to study the whole report. The style and vocabulary would therefore be different from the rest of the report, as it would be intended for "non-technical" readers. It can be a useful element in a **progress report**, for example, which may be written by engineers for financial managers; however, it should never be used as an alternative to an **abstract** in a thesis or academic work.
- Write it as a self-contained summary of the work, in about three to ten pages. As it is a mini-report in its own right, important figures and tables should be included—which you would not normally find in

a summary—and numbered separately from the main report. In the main report it would be incorrect to refer to something described in the executive summary; hence, it would be necessary to repeat these figures and tables.

F

Fair-use doctrine

- A limited amount of **copyright** material may be reproduced without infringing international copyright laws under the so-called fair-use (or fair-dealing) doctrine, which permits criticism, comment, review, or research of such material, provided that it is appropriately referenced.
- The amount of copyright material that may be reproduced legally—without permission from the copyright holder—is not defined explicitly, and it depends on the purpose and character of its use. There is greater leniency in the case of reports produced for academic rather than commercial purposes. It also depends on the amount and substantiality of the portion used in relation to the work as a whole.
- There are three aspects to this and you must consider all of them:

  (1) If you quote half a page from a five-page journal paper, that would not be regarded as fair use; however, if the original was a book of over 500 pages, that could be argued as fair use, as it is unlikely that your actions would have diminished the value of the original work.

  (2) If you transcribe something in its entirety (e.g., an essay or the entire discussion from a journal paper), irrespective of the amount, that is likely to be a problem.

  (3) If you do not add anything of substance—for example: by criticism, comment, or review—and simply reproduce other people's work, that is not fair use.

- The *Chicago Manual of Style* (University of Chicago Press, 1993, p. 146) has some practical guidelines, which include the following advice: "As a rule of thumb, one should never quote more than a few contiguous paragraphs or stanzas at a time or let the quotations, even if scattered, begin to overshadow the quoter's own material." Reproduction of copyright material beyond that deemed to be fair requires permission from the copyright holder, which may not be the author, but the publisher.

- Note that **figures** are not covered under the fair-use doctrine and permission is always required to copy a figure from someone else's work.
- See also **Plagiarism**.

Figures

- Figures are pictorial representations (e.g., charts, diagrams, flow-charts, photographs, maps). They are all referred to as figures and are never labeled *Diagram 1* or *Graph 1*, for example. A table is not a figure (so do not get caught out by calling it one). Computer programs (i.e., source code) and blocks of text (such as questionnaires and schematics) are also called figures.
- Never include a figure in your report without good reason: every figure should serve a purpose.
- Give each figure a caption (legend) and locate it below the figure. As readers tend to look at figures before reading the accompanying text, captions should *explain* what is in the figure. Captions are often written in sentence case (i.e., starting with a capital letter, followed by all words in lowercase, except for proper nouns, acronyms, and certain abbreviations), but that is only a matter of style preference. A period is not needed after the caption. Be consistent in the font size and style for all captions.
- Number all figures with standard Arabic numerals (i.e., 1, 2, 3, etc.) in the order in which they are first mentioned in the report (even if a more detailed discussion appears later). For a long report, it is advisable to use a two-part numbering system, where the first part designates the chapter and the second is allocated sequentially within the chapter—for example: *Figure 3.2* (or *Figure 3-2*) will be the second figure in *Chapter 3*, and *Figure B.1* (or *Figure B-1*) will be the first figure in *Appendix B*. If a figure comprises a series of

individual elements, identify each part as (a), (b), (c), etc. or (i), (ii), (iii), etc., and have a caption for each one.

- As figures are not covered under the **fair-use doctrine**, permission is always required to copy a figure from someone else's work. An example of a statement acknowledging such permission is illustrated in the figure below. Note that the term *courtesy of* is an acknowledgement used for material obtained free of charge and without restriction on its use. If you reproduce your own figure (say, from an earlier publication) you should still cite the original work; furthermore, it may be necessary to get the permission of the publisher to use the figure if they hold the copyright.

Figure 4.1 Test aircraft (reprinted, by permission, from Skywalker 2003)

- Always acknowledge the source of a figure that you were not personally responsible for producing (even when you have obtained the copyright holder's permission to use it). To copy someone else's figure without acknowledgement is **plagiarism**. This can be avoided by stating at the end of the caption: *from Skywalker (2003)*, for example, or alternatively: *after Skywalker (2003)*.

- Figures such as charts and diagrams that have been adapted, revised, or redrawn should be so indicated by writing after the caption: *Graph adapted from Phoolproof (2001a, p. 231)* or *Redrawn after Noonan et al. (2004, Fig. 3)*, for example. As copyright applies to the form of the chart or diagram (i.e., the way in which the data is presented) and not the data itself, permission is not required from the copyright owner to use a reformatted figure. However—and this is exceptionally important—failure to reference the source of the information is plagiarism.
- If you have many figures, prepare a list (see **List of figures/tables**). The facility in word-processing software to generate a list of figures, complete with page numbers, is especially useful; however, it does require a little planning. You can save yourself a lot of trouble by selecting a single **format style** for the captions and testing this feature early on in the write-up process.
- At an appropriate point in the text, draw the reader's attention to each figure (figures that are not referred to should be removed). Each figure should be placed in the report soon after reference is first made to the figure (see **Cross-referencing**).
- Figures that have a landscape orientation should be orientated such that the report has to be rotated clockwise to view the figure.

Font

For serious work choose a serious font style, such as Times, Times New Roman, Arial, Courier, or Helvetica; less formal styles—for example: Andy, Comic Sans—are better suited to e-mails. Use a 10 to 12 point font size. It is important to be consistent in font style and size for passages of text throughout the report; however, some authors like to reduce the font size (say from 12 to 10) for tables, captions, footnotes, long lists, references, and so on. Similarly, slightly larger font sizes are sometimes used for chapter and section headings. This is, of course, a matter of personal preference.

Footnotes and endnotes

- Footnotes and endnotes are useful when you want to clarify something or provide additional information, but cannot do so in the passage without disrupting the flow of the discussion.
- Superscript numerals—starting with 1 for the first note and numbered consecutively throughout the report—are used. As appendices are considered to be self-contained elements, notes in each appendix are usually numbered separately.
- *Footnotes* appear on the same page as the reference number, whereas *endnotes* are placed at the end of the section or chapter (in some rare cases they are located at the end of the report). Do not use both note types (i.e., footnotes and endnotes) in the same report. Footnotes are preferred as it takes the reader less time to find the information.
- Use abbreviations to condense the text in the note (see **Abbreviations (common)**).
- Notes intended to clarify **tables** or **figures** should not be numbered with text footnotes: use lowercase letters (i.e., a, b, c etc) or symbols (e.g., * † ‡ § ¶) to make this distinction if necessary. The sequence begins afresh for each table or figure.
- Footnotes and endnotes are also used for citing references in the humanities style; however, this is not recommended for scientific or technical work (see **Citing references (basic rules)**).

Formal writing

See **Style of writing (use of words)**.

Format (of report)

- Sort out the basic format of the report before you start writing; changing it afterwards can be a painful and unnecessary waste of time. You need to decide on the following:
  - Paper size (check that the word processor's default page set-up is correct).
  - Double sided or single sided print (this can influence the margins, position of page numbers, headers, footers, etc.).
  - Margins (print a test page, check the result and ensure that you have sufficient edge distance to bind the report).
  - Position of page numbers (see **Page numbers (pagination)**).
  - Headers and footers (these can be used for chapter numbers and/or headings; see **Headers and footers**).
  - Hierarchy (or structure) of headings for chapters, sections, and subsections; and heading styles (see **Headings**).
  - Text style (the **format style** defines the font type and size, line spacing, text alignment, and other text features).
- The above document formatting features can be stored as a template. Your organization may have such word-processing templates, but if you cannot get a suitable template, it is worthwhile creating your own, as this will ensure consistency of style throughout the report.
- Using unique pre-defined styles for chapter and section headings, figure captions, and table captions/titles will mean that the word-processing software will be able to generate a table of contents and lists of figures and tables for you, and update them to reflect any changes you make.

Format style

In the context of word-processing software, a *style* is a set of formatting characteristics that controls the appearance of the text—defining the font type and size; line spacing; text alignment; whether the text is bold, italic, or underlined; and so forth. Default styles, and ones that the user creates, are stored as templates and may be applied to new text.

See also **Format (of report)**.

Fowler's Modern English Usage

The *New Fowler's Modern English Usage* (rev. 3rd ed. by R. W. Burchfield, Oxford University Press, 2000) is widely regarded as one of the best reference books on English language usage, providing detailed advice on grammar, punctuation, style, and spelling (American and British English usage) and an astonishingly detailed treatment of troublesome words, with advice on how best to use them.

See also **Style manuals**.

Front cover

- The minimum information that should be put on the front cover of a *report* is
    - Title of report;
    - Report number (if appropriate);
    - Volume number and the total number of volumes (if appropriate);
    - Name(s) of the author(s) and qualification(s) (optional);
    - Organization(s); and
    - Date.

  It should also describe any restrictions on distribution, say for reasons of confidentiality.

- A typical *thesis* cover will indicate
    - Title of thesis;
    - Name of candidate;
    - Degree; and

- Year.

- Bound formal reports and theses usually have a title page (in addition to a front cover), which will provide the same information as that given above and a few additional facts (see **Title page**).

Future work

It is a good idea to identify the next steps to be taken to progress the work beyond the point that you have brought it to. Avoid vague or simplistic statements, such as: *This requires further investigation.* Be specific and indicate the direction and approach that should be adopted. This advice to the readers is usually presented in the **recommendations**; alternatively, if there are just a few suggestions that you wish to make, they can be included in the **discussion**.

G

Getting started

For most engineers, report writing is as appealing as a visit to the dentist—and the process can be equally painful. The best approach is to break up the work into manageable chunks and get going as early as possible. It does not matter which section you start with—pick any section that you feel comfortable with.

Set out a rough order of the chapters, and expect to revise it several times as the report develops. Write rough notes for individual chapters, and sketch out as much of the report as you can, without spending too much time on the details. Do not agonize over the exact wording of every sentence—just keep on writing. When you come back to it later, fresh ideas will emerge on how to rephrase sentences and re-organize the material. Focus on the primary **objectives** of the work and describe the most important elements first; this will aid in getting the report structure correct (see **Structure of reports and thesis**). Most authors maintain that it gets easier with practice; this may be true, but it is never going to be fun.

Develop the reference list as you go along—never leave it to the last moment. Keeping track of cited references when using the numeric method is time-consuming and it is easy to make a mistake—so consider using appropriate software for this purpose. For a large report with lots of references, it is relatively easy to use the author-date referencing method and keep updating an alphabetical list of references (with *complete* details). If you do not have a software referencing program and plan on using the numeric method, consider writing the report with the author-date method and then changing over later when you are nearly finished (see **Citing references (basic rules)**).

See also **Format (of report)**.

Gimmicks, clip art, and emoticons (smileys)

- Gimmicks and clip art—pasted in to brighten up the work or to provide light-hearted relief—just do not work: leave them out of your report.
- A flashy little banner, such as the one reproduced below from the dedication of a student report, is tacky and should be avoided.

- The same advice goes for little airplane silhouettes and smiley icons, like these: ☹ ☺ , which are occasionally used for bullets.
- Emoticons (which is the technical word for what everyone calls smileys), for example: :-) :-(;-) :-o =) , have lost their novelty in e-mails and have no place in formal reports.

Glossary

A glossary is a list of the less well-known terms that have been used, with short explanations (the example that follows is a glossary of grammatical and other terms used in this book). It is a good idea to include a glossary in a technical or scientific report written for an audience who would not be familiar with the terminology used or a report with lots of foreign words.

Locate the glossary after the **table of contents** or, alternatively, at the start or end of the **appendix** (i.e., where it is easy to find). If the explanation for a particular entry runs over one line, indent subsequent lines of the entry (i.e., a hanging indent).

Glossary of terms used in this book

adjective A word describing or qualifying a noun (e.g., immense, purple, sour, predatory, definitive).

adverb A word that alters the meaning of an adjective, verb, or another adverb, typically by expressing time, manner, place, or degree (e.g., immediately, very, here, softly, harshly). An adverb can also modify an entire sentence.

antecedent The noun or noun phrase referred to by a pronoun (e.g., Ice particles [*antecedent*] grow by aggregation as they [*pronoun*] fall within clouds).

appositive A noun or noun phrase that renames, identifies, or defines the noun or noun phrase immediately preceding it (e.g., Dr. Binoche, *Technical Director of N-vent Enterprises*, patented the concept).

Arabic numerals Number symbols: 0, 1, 2, 3, etc.

article The words *a* and *an* (indefinite articles) and *the* (definite article), which precede a noun.

clause A segment of a sentence containing both a subject and a verb—or a group of words containing a verb—that expresses what the subject is or does (e.g., *The salmon migrate* in late spring).

colloquialism An informal word or expression (e.g., bummer, math, lab, kick the bucket).

conjunction A word used to join words, clauses, or sentences (e.g., and, but, because, while, however).

contraction A shortened form of a word or group of words (e.g., gov't, can't, it's, who's).

euphemism A polite word or phrase used instead of a harsh, unpleasant, or embarrassing term (e.g., to sleep with).

idiom A figure of speech, the meaning of which can not be deduced from the actual words (e.g., to leave no stone unturned).

independent clause A clause that can stand alone as a complete sentence (e.g., *An albatross is a nomadic bird* that can cover great distances over the oceans).

italic typeface The sloping style of typeface or printing (e.g., *This is an example of italics*).

modifier A word (typically an adjective or noun) that qualifies or restricts the meaning or application of another word or phrase (e.g., There was a *partial* solar eclipse).

noun A name of a person, place, or thing; or class of people, places, or things (e.g., Charlie, Amsterdam, pen, Chinese, towns, animals). A proper noun is the name of a particular person, place, or thing (e.g., Juan, Antarctica, Mississippi) and has a capital letter.

object The noun, noun phrase, or pronoun in a sentence or clause to which the verb acts (e.g., The X-rays caused the *hydrogen molecules* to vibrate).

ordinal numbers Numbers that designate position in a series (e.g., first, second, twelfth).

parenthesis A word, clause, or sentence inserted into a passage as an explanation, elaboration, or afterthought, marked off by commas, dashes, or parentheses (e.g., The test case, *developed by the author*, was evaluated twice).

person Category of pronouns, possessive determiners, and verb forms based on whether reference is made to the speaker (first person), the addressee (second person), or a third party (third person).

preposition A word, usually preceding a noun or pronoun, that expresses a relation to another word or clause (e.g., The spider *on* the wall moves *past* the light switch).

pronoun A word that can function as a noun, making reference to someone or something (e.g., I, you, we, he, she, it, this, that, them). The first-person pronouns are *I* and *we*.

rhetorical question A question asked, not to elicit information, but rather to make a statement (frequently followed by a "response" from the inquirer).

Roman numerals Number symbols: I, II, III, etc. (uppercase) and i, ii, iii, etc. (lowercase).

roman typeface The ordinary style of typeface (print) with small upright letters, as used in this sentence.

subject The noun, noun phrase, or pronoun which indicates what the sentence or clause is about (e.g., The *spacecraft* has a mass of 40 kg).

tense The form or state of the verb, such as *past*, *present*, or *future*, that indicates the time of the action of the verb relative to the time of speaking.

verb A word describing an action, state, or occurrence (e.g., hit, evaluate, compute, write, infer).

Grammar and style

Grammar: it can be one of the most boring subjects in the world. To be coached on the peculiarities of English grammar is a horrible fate for many technically-minded people who, at the mention of the words *split infinitive*, would make a dash for the door. Unfortunately, poor grammar distracts from good report-writing, so you need to make a concerted effort (no matter how painful) to stay focused and correct the serious blunders. Thirteen problem areas are described:

(1) Double negatives (i.e., negative terms that can reinforce or cancel each other)

(2) Incomplete sentences

(3) Misplaced adverbs

(4) Mixing up *a* and *an* before words and acronyms

(5) Old-fashioned teaching (e.g., not splitting infinitives or starting sentences with conjunctions or ending them with prepositions)

(6) Parallelism (i.e., using similar grammatical form to express ideas of similar content)
(7) Plurals (of irregular nouns, symbols and acronyms)
(8) Possessives (using an apostrophe)
(9) Pronouns (problems with usage)
(10) Restrictive and non-restrictive clauses (e.g., *that/which* confusion and using appositives)
(11) Spelling and incorrect word usage
(12) Subject–verb agreement (e.g., confusing *is* with *are*)
(13) *Who* or *whom*: which is correct?

Grammar: (1) Double negatives

- In colloquial English, a second negative term in a sentence is used to reinforce the first (e.g., They didn't see nothing), whereas in standard English the second term effectively cancels the first (changing the meaning to: *They saw something*). In formal report-writing neither colloquialisms nor potentially ambiguous statements should be used. Double negative constructions of this type should thus be avoided.

- It is advisable to use positive statements whenever possible—for example, instead of saying: *It is unlikely that this instruction will not be misunderstood*, you could say: *It is unlikely that this instruction will be understood*, or better still, *It is likely that this instruction will be misunderstood.*

Grammar: (2) Incomplete sentences

- The omission of an element (e.g., a word or clause) from a sentence can render it meaningless—for example: *The administration of a placebo, which could have a positive effect not attributed to any physiological or chemical change, but due to a psychological benefit associated with the patient's belief in the treatment.* This appears to tell us something about the way in which a placebo works—that is, it has "*a positive effect not attributed to any physiological or chemical change* ...”—but that part of the sentence is only a restrictive clause that should introduce the main part (i.e., concerning the administration of a placebo), and that main part is missing.

- A related problem occurs when a fragment of a sentence hangs on to the previous sentence in an informal construction. The second part of each of the examples below does not meet the requirements for a sentence in formal writing. In an informal context this would not be a problem, but in formal writing it would be better if each pair of sentences were revised either by joining the two sentences or by rewording the second sentence.

  Examples:

  - The market price for polished gemstones had remained high. The reason being that a cartel controlled the supply of uncut stones.
  - It was noted that safety standards in the mine had deteriorated over the past 12 months. Because of longer working hours (due to reduced staff levels), repeated flooding of the lower chamber and the poor reliability of Doc’s new drilling machine.
  - The seven partners presented strategies to revive their ailing business. For example: digging a new shaft, purchasing pneumatic drills and sending Mr. Grumpy to Mining College.
  - Extensive mining had led to reduced yields. Which is why Seven Dwarfs Inc. diversified into diamond cutting and polishing.

Grammar: (3) Misplaced adverbs

- Adverbs can act as modifiers in a sentence: qualifying or restricting the meaning of a word or phrase. However, adverbs placed in the wrong position in a sentence can lead to ambiguous statements or awkward phrasing. Examples of adverbs that can be incorrectly placed are: accurately, almost, chiefly, completely, critically, extremely, frequently, fully, incorrectly, moderately, mostly, nearly, not, often, only, practically, really, totally, very, wholly.

- Adverbs are best placed adjacent (or very close) to the word or phrase they modify. Consider the following examples:

  - *Inspecting the part frequently leads to repairs.* Now, *frequently* can refer to *inspecting* (i.e., frequent inspections) or to *repairs* (i.e., repairs often result). If it is *frequent inspections* that is intended then the sentence should be reworded to avoid confusion: *Frequent inspections of the part leads to repairs.*

  - *The designers only produced three prototypes.* This could, possibly, mean that only the *designers* (not anyone else) produced the prototypes, or it could mean that the designers only *produced* (and did not, for example, test) the prototypes, or it could mean that only *three* prototypes were produced (not four or more). If the issue is the number of prototypes, then it is best to say: *The designers produced only three prototypes.*

  - *The design was intended to not represent an optimized solution.* This is an awkward construction, which can be improved by revising the order of the words as follows: *The design was not intended to represent an optimized solution.*

Grammar: (4) Mixing up *a* and *an*

The general rule is that *a* precedes words that start with a consonant (e.g., a book, a canister, a duck) and *an* precedes words that start with a vowel (e.g., an article, an exercise, an illusion, an observer, an umbrella), with the following exceptions:

(1) Before words and acronyms that start with a long vowel *u* (sounds like the letter U) (e.g., a unit, a utility, a U-turn, a URL);

(2) Before words starting with a silent *h* (e.g., an honest, an hour); and

(3) Before abbreviations that are pronounced as individual letters (i.e., initialisms) that *sound* as though they start with a vowel

(e.g., an FDA [directive], an HB [pencil], an LCD, an MBA, an NDT, an R&D, an SGML).

Grammar: (5) Old-fashioned teaching

- There are a few contentious rules of grammar, which were taught to previous generations of schoolchildren, that are today regarded as nothing more than preferences of style—for example:
  (1) Never split an infinitive.
  (2) Never start a sentence with a conjunction (e.g., and, but).
  (3) Never end a sentence with a preposition (e.g., above, in, over, with).
- While readers schooled in traditional English usage will raise their eyebrows (or worse, make sniggering comments about illiteracy) when they spot these "mistakes," there is, in fact, nothing wrong with sentences that violate these conventions.
- Further details on these three "rules" are given below:
  (1) To split an infinitive is to write something between the word *to* and the verb in a sentence—for example: *It was decided to immediately stop the experiment.* This, it would have been argued, splits the infinitive *to stop* and the remedial action would have been to relocate the word *immediately.* However, avoiding split infinitives can sometimes lead to cumbersome or ambiguous sentences. Consider two alternatives: *It was immediately decided to stop the experiment*, and *It was decided to stop the experiment immediately.* Note that the first version has a different meaning to the original.

  (2) Starting a sentence with a conjunction—in particular the word *and*—is a powerful tool that can be used to link two complex sentences that would otherwise be difficult to understand if written as a single sentence. Overuse of this construction, though, especially in a rambling, unstructured brain dump, is to be avoided: not because it is bad grammar, just bad writing.

(3) Sentences ending in a preposition, such as: *It was a thermocouple, not a thermometer, that he measured the temperature with*, can always be reworked: *It was a thermocouple, not a thermometer, with which he measured the temperature.* Both versions are correct, but the first may not sound right to all readers; it is best to avoid such constructions in formal writing.

Grammar: (6) Parallelism

- If a series of related sentences is constructed with a common format (i.e., with the same tense, verb position, style, and so forth), a parallel structure in the composition is created, which greatly enhances its readability. This principle of parallelism can be summed up as follows: *similar content is best expressed using a similar form.*

- The principle can be applied to lists of instructions or to describe procedures. As an illustration, consider the following extract from a pilot's flight manual:

  (1) Raise the nose wheel at 32 kt.
  (2) Rotate and lift-off at 54 kt.
  (3) Accelerate and climb-out at 60 kt.
  (4) At 300 ft, you should retract the flaps.
  (5) At 400 ft, the power is to be reduced to 2450 rpm.

  Now, the simple pattern established by items 1–3 is suddenly broken by item 4 when the pronoun *you* is used and the verb (retract) is not placed at the start of the sentence. And then, in item 5, the verb is written as *reduced*, whereas to conform to the pattern it should be *reduce*. It would be better to rewrite the last two instructions as

  (4) Retract the flaps at 300 ft.
  (5) Reduce the power to 2450 rpm at 400 ft.

- The principle can also apply to a single sentence—for example: *The transportation working group set as goals for new vehicles an ambitious 50% cut in fuel consumption and a reduction in* NO_x

(nitrogen oxides) of 80%. This could be rewritten as *The transportation working group set as goals for new vehicles an ambitious 50% cut in fuel consumption and an 80% cut in* NO_x *(nitrogen oxides).* Note that by repeating the word *cut* the sentence is not weakened but actually strengthened.

- ❑ If a series of adjectives is written in a consistent manner, the sentence is easier to read—for example, the statement: *The new alloy is lighter, stronger and is less expensive than the original material*, can be reworked as *The new alloy is lighter, stronger and cheaper than the original material.*

- ❑ Similarly, articles (e.g., a, an, the) and prepositions (e.g., in, on, at, to) associated with items in a list or series are best treated in a consistent manner. When an article or a preposition applies to all items in a list or series, it should appear only before the first item or before each item. For example, do not write: *The image recognition software identified the red sphere, blue cube, and the orange disk*; rather *The image recognition software identified the red sphere, blue cube, and orange disk.* An even better version is *The image recognition software identified the red sphere, the blue cube, and the orange disk.*

- ❑ Clauses in a multi-part sentence linked with *either* or *neither* or *both* should be parallel in form. Consider the following examples:

  - *Flight plans must be submitted for all cross-border flights: pilots may either phone the service desk or hand in a completed form to the control tower or details of the flight pilots can supply by radio once airborne.* In this sentence the word *either* has a verb (i.e., phone) after it, leading the readers to expect another verb after *or.* Now, after the first *or* there is *hand in* (which is okay), but after the second *or* there is a noun (i.e., details). It is better to rephrase the last part: … *or supply the details of the flight by radio once airborne.*

  - When a preposition (e.g., by, in, on, to) follows *either* it is best repeated after *or*—for example: *A nuclear chain reaction may be stopped using boron (which absorbs neutrons), either by dropping boron rods into the reactor core or by increasing the boron concentration in the coolant.*

Grammar: (7) Plurals

- The general rule is to add an *s* (e.g., experiments, specimens) or *es* (e.g., classes, echoes) to form a plural. Nouns that end in *y* following a consonant change to *ies* (e.g., policies, studies) and most nouns that end in *f* or *fe* change to *ves* (e.g., halves, knives), but not all (e.g., beliefs, proofs).
- There are many exceptions to the general rule in science and engineering:
    - Be careful with irregular plural forms—for example: analysis/analyses, appendix/appendices, axis/axes, bacterium/bacteria, colloquium/colloquia, criterion/criteria, genus/genera, hypothesis/hypotheses, larva/larvae, locus/loci, matrix/matrices, medium/media, nucleus/nuclei, ovum/ova, phenomenon/phenomena, quantum/quanta, radius/radii, stimulus/stimuli, stratum/strata, symposium/symposia, vertebra/vertebrae, vortex/vortices (singular form first, followed by plural). If in doubt, check in a dictionary.
    - There are a few plural nouns that are unchanged from the singular form (e.g., aircraft, chassis, deer, fish, sheep).
    - There are also a small number of abstract nouns (e.g., darkness, honesty, wisdom) that have no plural form and the odd plural noun that has no singular form (e.g., goods, species).
- An article consisting of two similar parts that is described as a *pair of* (e.g., binoculars, clippers, gloves, pincers, pliers, scissors, shears, tweezers) takes the singular form (and requires a singular verb—for example: a pair of scissors *was* removed). When the words *a pair of* is omitted, it takes the plural form (e.g., scissors *were* removed).
- An apostrophe should not be used with people's names to indicate a plural—for example: *The work undertaken by the Newman's concerning micro-fluidics*. It should be … *by the Newmans* ….

- The plurals of dates, numbers, letters, and abbreviations—as a general rule—do not have apostrophes (e.g., the 1990s, seven 4s, CFCs). Exceptions occur with plurals of single letters and symbols, where to eliminate confusion an apostrophe is sometimes used (e.g., A's, a's, K_n's). If the plural of a term such as K_n is likely to be misunderstood, the sentence should be rewritten using a phrase such as: *values of* K_n or *various values of* K_n or *values for various* K_n.

- Special rules apply to the plurals of abbreviations (e.g., *pp.* for pages, *vols.* for volumes). See **Abbreviations (common)**.

- In Latin, *data* is the plural of *datum*; in traditional English usage and in certain scientific disciplines, it is also treated as plural. So to be absolutely correct you should use plural verbs—for example: instead of writing *the data is indicative of ...*, you should write *the data are indicative of ....* However, this can sound clumsy and in modern English it is acceptable to use *data* as a singular collective noun (such as *information*).

Grammar: (8) Possessives

- The general rule is to add an apostrophe and an *s* to a singular noun to give the possessive form (e.g., plant's foliage, Havana's proximity, Nyman's postulation) and just an apostrophe to a regular plural noun (e.g., students' results, insects' nest, Ethiopians' abilities). Plural nouns that do not end in an *s* take an apostrophe and an *s* (e.g., men's body mass, children's interests).

- The possessive of a compound noun is formed by adding *'s* to the end of the word (e.g., double-decker's height, self-tester's response); similarly, *Eremenko and Wu's formulation* has *'s* added only after the second name.

- The possessive form of proper nouns (in particular personal names) that end with an *s* or *z* sound is problematic (largely due to differing opinion on what is correct). A pragmatic rule (but not one that is accepted by all grammarians) is: "If it ends with an *s* sound, treat it as singular" (e.g., Stokes's, St. Francis's, Thomas's, Loftus's) and "if it ends with a *z* sound, treat it as plural" (e.g., Williams', Daniels', Evans', Adams'). Similarly, names ending in an *eez* sound and ancient classical names are treated as plural (e.g., Archimedes', Ramses').

- Most pronouns change form to indicate a possessive (e.g., *it*, *he*, *her*, *you*, *we*, and *their* become *its*, *his*, *hers*, *yours*, *ours*, and *theirs* respectively). These pronouns do not take an apostrophe (e.g., the flower lost *its* petals; the results were *hers*; the discovery was *theirs*). Note that *it's* is a contraction meaning *it is* or *it has* and does not indicate a possessive. A few pronouns do not change form and these require an apostrophe (e.g., *one's* approach to testing; it was *everyone's* responsibility).

Grammar: (9) Pronouns (problems with usage)

- A pronoun is a word that can take the place of a noun, making reference to someone or something (the someone or something is called the antecedent). First-person pronouns refer to the speaker or speakers (e.g., I, we, me, us, my, our, mine, our, ours, myself, ourselves); second-person pronouns refer to the person or persons being spoken to (e.g., you, your, yours, yourself, yourselves) and third-person pronouns refer to someone else (e.g., him, her, them, their, his, hers, theirs, himself, herself, themselves). Pronouns can also refer to something (e.g., it, its, itself, this, that, those).
- There are three problem areas:
  - (1) Avoiding the use of first-person pronouns,
  - (2) Ensuring that pronouns match their antecedents in gender and number, and
  - (3) Ensuring that antecedents are obvious to the readers.

(1) *Avoiding first-person pronouns*

It is not for grammatical reasons that first-person pronouns are to be avoided, just that reports are, by their nature, impersonal, and by avoiding these pronouns, the personality of the writer is hidden (which is generally considered to be a good idea). For example, the statement *I analyzed the data*, can be rewritten as *The author analyzed the data.* Alternatively, the **passive voice** can be employed: *The data was analyzed [by me].* Now, the *by me* is almost always omitted (to comply with the custom of not using first-person pronouns). This is seldom a problem as readers will automatically assume that the writer performed the action; however, if other people are mentioned in the passage (say, in a citation) then the sentence could be ambiguous.

(2) *Matching pronouns with their antecedents*

A pronoun must always match its antecedent (the item(s) or person(s) to which the pronoun refers) in gender and number—for example:

- *Ann-Marie Willis and her colleagues published the paper.* Matching gender is straightforward if the person is named.
- *Willis et al. (2003) published the results of their study.* There is more than one person referred to and hence a plural pronoun is needed.
- *The operator is required to put his foot firmly on the red pedal.* There are a few ways of avoiding the sexist *his* in sentences of this type: replace *his* with (1) *his or her*, or (2) *his/her* or (3) *their*. The first two options are clear-cut, but the third is controversial as many people would perceive a mismatch between the noun *operator* (which is singular) and the pronoun *their.* Rewriting such sentences in the plural form can sometimes solve the problem, but in this particular case, the revised version: *The operators are required to put their feet firmly on the red pedal* has a different meaning to the original.
- Singular antecedents linked by *or* will require a singular pronoun—for example: *If there is a slot or hole in the outer surface, it will allow the air to escape.* Similarly, plural antecedents (or antecedents linked by *and*) will require a plural pronoun—for example: *If there are slots or holes in the outer surface, they will allow the air to escape.*

(3) *Making antecedents obvious to the reader*

Examples:

- *Berkman (2002) reproduced the results of Wildman (1995) using his test equipment.* Whose equipment was used: Berkman's or Wildman's? It is not obvious.
- *The streamline curvature causes an imbalance between the pressure forces and centrifugal forces in the boundary layer, which in turn creates a flow velocity normal to the local streamline. This has a point of inflection, which makes it intrinsically unstable.* The second sentence is a mess: the pronouns *this* and *it* refer to something in the first sentence, but it is not obvious what they refer to; it would be better to rewrite this sentence repeating the antecedents.

Grammar: (10) Restrictive and non-restrictive clauses

(1) *What is a restrictive clause?*

- A *restrictive* (or defining) clause provides essential information about the noun or phrase preceding it; the clause cannot be omitted and the sentence cannot be rewritten as two sentences without a change of meaning. For example: *The cave that opens to the sea floods in winter.*
- A *non-restrictive* clause provides additional information about the noun or phrase preceding it, but does not define, limit, or identify it. Such a clause is parenthetical (i.e., an insert into a sentence, typically marked by a pair of commas); it interrupts the main flow of the sentence to explain or elaborate on something and can be omitted to leave a less informative, but grammatically complete, sentence. For example: *The Hadar Valley, where the hominid skeleton Lucy was found, is a barren wasteland in Ethiopia.*

(2) *That/which confusion*

- *That* can be used to introduce a restrictive clause in a sentence (there is no comma before the word *that*, when it is used in this way). *Which* can introduce a non-restrictive clause (and there is a comma before and after the clause).

  Examples:

  - *The new keyboards that are faulty are in the workshop.* This identifies the keyboards, that is, the faulty ones, and they are all in the workshop. It is implied that the good ones are elsewhere.
  - *The new keyboards, which are faulty, are in the workshop.* The parenthetical clause *which are faulty* tells us something about the keyboards, but does not identify, or define, them—all the new keyboards are faulty and they are in the workshop.

- The distinction between restrictive and non-restrictive clauses is important, especially in technical writing where ambiguous statements can have serious consequences. Using *which* only for non-restrictive clauses improves clarity; but despite the recommendation of this practice by many authorities, it is not universally observed, and *which* also gets used for restrictive clauses—hence the confusion. The rule to remember is: "*that* defines, *which* informs."

(3) *Appositives*

- An appositive is a grammatical term for a noun or noun phrase that serves to add information by renaming, identifying, or defining the noun or noun phrase immediately preceding it.

  Examples:

  - *The Lammergeier, or Bearded Vulture, has a wingspan of over 8.5 ft.*
  - *The highest mountain range in Southern Africa, the Drakensberg, was the selected location to study the birds.*

- An appositive can also be used to define an abbreviation, acronym, mnemonic or little-known term—for example: *An unmanned air vehicle, UAV, was used to film the birds' wing motions during flight.*

- In most cases, appositives are non-restrictive (and marked by a pair of commas), but they can also be restrictive (and commas are not required).

    - Non-restrictive appositive: *Dr. Peter Wolfe, the well-known ornithologist, presented his research.* Here the appositive *the well-known ornithologist* is non-restrictive; it provides extra, but non-essential, information about Dr. Wolfe (without the appositive readers would still know that it was Dr. Wolfe who had presented the research).

    - Restrictive appositive: *The well-known ornithologist Dr. Peter Wolfe presented his research.* Here the appositive *Dr. Peter Wolfe* provides essential information that defines (or, in this case, *identifies*) the ornithologist (without the appositive readers would not know who had presented the research).

Grammar: (11) Spelling and incorrect word usage

- Mixing up similar or identical-sounding words (homophones) (e.g., *principal* and *principle*) can have amusing consequences; in most cases, it just appears as poor editing (with the reader working out what was meant); however, in some instances, it can totally change the meaning of the sentence. For example: *The chairperson stated that the response to the criticism would be countered by three discreet actions* (or should that be *discrete*?). Most word processors come with a built-in thesaurus: use it, but with care!

- See **Incorrect word usage (confusables)** for a list of words that are frequently mixed up. See also **Spelling** and **Hyphenation and word division**.

Grammar: (12) Subject–verb agreement

- What is the correct form of the verb? Do you use *is* (singular) or *are* (plural), *has* (singular) or *have* (plural), *was* (singular) or *were* (plural)? The rule is simple enough: if the subject is singular, then the verb is singular, and if it is plural, then the verb is plural. The problem is to identify the subject.

- Examples:
  - *A black and white or colored print is acceptable.* A singular verb is required after a list of singular nouns separated by *or*. (A *black and white print* is singular).
  - *Distance and acceleration are to be measured.* When nouns are joined by *and* the subject is plural.
  - *The speed at which insects will rupture on impact is related to their mass*. The subject is *speed* (i.e., singular).
  - *One of those techniques that are easy to master involves ....* When a relative clause (i.e., a clause introduced by a relative pronoun, such as *that* or *who*) follows an expression of the form "*one of those/the ...*" it requires a plural verb.
  - *The control and monitoring system has a built-in-test function.* The subject is the *system* (i.e., singular).
  - *Pliers have been used.* Pliers (like clippers, pincers, scissors, shears) require plural verbs (and plural pronouns); however, the phrase *a pair of pliers* (or clippers, pincers, etc.) requires a singular verb—for example: *a pair of pliers has been used.*
  - *Neither test was successful.* The singular verb form is required after the word *neither* (or *either*). Here it refers to a *test*, which is singular.
  - *Sensitivity studies based on the selected design point were conducted.* The subject is *studies* (i.e., plural).
  - *The first specimen in each of the four batches was discarded.* The subject is a single *specimen.*

- *Fifteen percent of the answers are wrong.* When *percent* refers to something that may be quantified by number (e.g., *answers*) rather than quantity (e.g., *oxygen*), it takes a plural verb form.
- *Six minutes is insufficient.* Measurements and phrases that quantify something take singular verb forms.

❑ There are a few special cases:

- When the phrase *a number of* is followed by a plural noun, it takes the plural form (e.g., there *are* a number of solutions to this problem; a number of researchers *have* addressed this problem). However, if the phrase *the number of* is followed by a plural noun, it takes the singular form (e.g., the number of tests to be conducted *is* small).
- Organizations are treated as singular, even if the name appears plural (e.g., Short Brothers of Belfast *is* recruiting).
- A collective noun denoting a number of individuals (e.g., crew, crowd, group, herd, staff, team) can be followed by either a singular or a plural verb, depending on whether the subject is considered a single unit or individuals (making up a group). For example: *the submarine crew is small* (i.e., few members); *the submarine crew are small* (i.e., small individuals). Collective nouns describing inanimate objects (e.g., baggage, collection, fleet, forest, set) are always regarded as singular.

Grammar: (13) *Who* or *whom*: which is correct?

❑ *Whom*—the dictionary will tell you—is the object form of *who*. But that is not particularly useful. There is, however, a simple way of checking which is correct:

Step 1. Extract the clause containing the word *who* or *whom*.

Step 2. Rewrite the clause replacing *who/whom* with either (a) *he*, *she*, or *they*; or (b) *him*, *her*, or *them*.

Step 3. If the clause sounds right with *he*, *she*, or *they*, then *who* is correct; if it sounds right with *him*, *her*, or *them*, then *whom* is correct.

- Examples:
  - *The editorial board named the researchers who/whom it said had made a significant contribution to the biomedical field.*

    The clause in question can be rewritten as: *it [the editorial board] said they made a significant contribution ….*

    Conclusion: *who* is correct.

  - *In his opening statement, Kheiber (2004) paid tribute to Brunel, who/whom he admired.*

    The clause can be written as: *he [Kheiber] admired him.*

    Conclusion: *whom* is correct.

  - *Litsa Papadopoulos criticized Mario Valenzuela (2005), who/whom she thought was consistently wrong.*

    The clause can be written as: *she [Litsa Papadopoulos] thought he was consistently wrong.*

    Conclusion: *who* is correct.

  - *The man with who/whom Charles Fairweather conducted his pioneering studies died in 2002.*

    The clause can be written as: *Charles Fairweather conducted his pioneering studies with him.*

    Conclusion: *whom* is correct.

Graphs

See **Charts**.

Greek alphabet

Greek alphabet: Upper- and lowercase letters

Name	Uppercase	Lowercase
alpha	Α	α
beta	Β	β
gamma	Γ	γ
delta	Δ	δ
epsilon	Ε	ε
zeta	Ζ	ζ
eta	Η	η
theta	Θ	θ
iota	Ι	ι
kappa	Κ	κ
lambda	Λ	λ
mu	Μ	μ
nu	Ν	ν
xi	Ξ	ξ
omicron	Ο	ο
pi	Π	π
rho	Ρ	ρ
sigma	Σ	σ
tau	Τ	τ
upsilon	Υ	υ
phi	Φ	φ
chi	Χ	χ
psi	Ψ	ψ
omega	Ω	ω

H

Hand-drawn sketches

Widespread availability of computers means that readers no longer have to decipher handwriting that resembles a spider's waltz across the page. One downside of this positive development is the misguided perception that if something is not computer-generated it will not be acceptable in a report. Hand-drawn sketches and drawings are an important feature in many technical disciplines; however, in line with the unnecessary trend to make technical reports appear as published booklets, there is a tendency to leave sketches out altogether or to replace them with poor quality computer-drawn graphics. This is silly as it entirely misses the point: reports are written to convey facts and ideas to readers. And neat hand-drawn sketches are often the best and fastest way to record many graphical ideas and concepts. You can always use a scanner and import the image into the report if an electronic version is required.

Sketches are treated as **figures**—give each one a figure number and a caption (which should be located below the sketch).

Harvard system

This is the author-date referencing method. See **Citing references (examples of author-date method)** and **References (examples of author-date method)**.

Headers and footers

- Headers and footers provide useful information for the reader to navigate the report. They are also an important part of **revision control** of technical documents. In addition to the page number, which is always required, it may be desirable to include
    - Organization name;
    - Shortened report title or report number or both;

- Shortened chapter title or chapter number or both;
- Section author or responsible person's name; and
- Revision number or letter.

- Company technical *reports* often have a prescribed header and/or footer format providing specific details, as illustrated in the following engineering report header:

THREE LITTLE PIGS CONSTRUCTION INC. BACKWATER, LA		*REPORT* *NO. TR-167*
Compiled by: *Paul Pig*	Date: *September 16, 2004*	Page 24 of 41
Checked by: *Brian Pig*	Date: *September 18, 2004*	

- *Theses* tend to have a format specified by the university; most would have only the page number at the top or bottom of each page, but increasingly there is a trend to provide a chapter number and/or a chapter title in the header.

Headings

- Avoid ambiguous headings: keep them simple and descriptive.
- Use capitalization, underlining, font size, and typeface (i.e., the **format style**) to distinguish between different levels of headings. Select font sizes and styles that result in the major headings standing out from the surrounding text and the subheadings being less noticeable. A period is not needed after the heading.
- Be meticulously consistent in implementing the selected heading style—all headings at the same level in the report must have the same format and position with respect to the edge of the page (i.e., flush left, indented, or center). If you use "sentence case" (i.e., starting each heading with a capital letter, followed by all words in lowercase, except for acronyms and proper nouns), do so consistently.

- Set up the heading structure, including format styles and numbering system, at the start of the write-up process. Word-processing software can automate the construction of a **table of contents** if the headings are correctly identified. Use it, not only to save time, but also to check the consistency of the headings.
- Use standard Arabic numerals (i.e., not Roman numerals or Greek letters) to identify the headings—see **Numbering system (for chapters and sections)** and the example below:

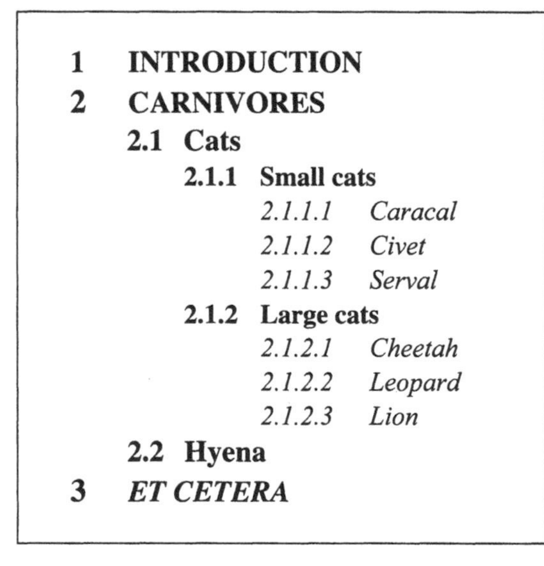

1 INTRODUCTION
2 CARNIVORES
2.1 Cats
2.1.1 Small cats
2.1.1.1 Caracal
2.1.1.2 Civet
2.1.1.3 Serval
2.1.2 Large cats
2.1.2.1 Cheetah
2.1.2.2 Leopard
2.1.2.3 Lion
2.2 Hyena
3 *ET CETERA*

- Avoid writing bits of introductory text between a heading and a subheading (i.e., between *2 Carnivores* and *2.1 Cats*, in the example above). If you need to write something at that point of the report insert another heading at the same level as the subheading (readers like to be able to refer to elements of a report by unique chapter or section numbers).
- Do not have only one subheading at the lowest level in any one subsection. For example, if there was only one level four heading under "2.1.1 Small cats" (say the *Caracal*), change the heading of

subsection 2.1.1 to include the subject of the lower heading and remove the fourth heading level for that subsection.

- Never put a heading at the bottom of a page without any text on that page.
- Readers find it difficult to cope with a heading structure of five or more levels, so avoid numbering a subsection *2.1.2.3.1*, for example. At that level it is best to use *(a)*, *(b)*, *(c)*, etc. or *(1)*, *(2)*, *(3)*, etc. to distinguish between different elements in the report.
- See also **Structure of reports and theses**.

Honesty

Be honest in your report writing: never exaggerate or falsify results. Reflect openly on the shortcomings and limitations of the work. Good reports can be based on seemingly "bad" results if you have an in-depth understanding of what happened. It is even better if you can indicate how to get it right next time.

See **Bad results**.

Humor

The **introduction** and **discussion** can be written in an engaging and interesting way, but keep jokes, irony, clever plays on words, witty remarks, and so forth out of your report: something amusing in one culture can be confusing—or even offensive—in another culture or language.

Hyphenation and word division

Hyphens join words together (e.g., self-test, mid-1970s, center-to-center) and add prefixes to words (e.g., pre-existing, sub-Sahara)—with the specific purpose of preventing misunderstanding (a *cross-section of the graduating class* is not the same thing as a *cross section of the graduating class*).

There are a lot of rules defining the use of hyphens, but—and this is the problem—there are also many differences of opinion (three editors are likely to give you four viewpoints). Dictionaries do not always agree and words such as *benchmark* and *takeoff* can be written with or without a hyphen. There is also a tendency to drop the hyphen with familiarity and to write the word "closed up" (a trend accelerated by the Internet); so, words that were hyphenated several years ago may not be written that way today (e.g., lifestyle, website). To make matters even more complicated, there are some differences between American and British English (there is a greater acceptance of hyphenated words being written as one in America).

Hyphenation: general guidelines

- A compound phrase used as an adjective before a noun should be hyphenated (e.g., long-term plan, spur-of-the-moment decision, large-scale facility, matter-of-fact approach, energy-intensive fuel, low-density material, six-person committee). Similarly, a compound phrase used as a verb is hyphenated (e.g., to beta-test, it was beta-tested). When used as neither an adjective nor a verb, many of these word combinations do not require a hyphen (e.g., short-term strategy, in the short term; the center-to-center distance, measured from center to center).

- A hyphen is not required when an adjective is modified by an adverb (e.g., clearly visible flaw). In this case (unlike the examples above), the adjective *visible* can stand on its own before the noun *flaw* (additional examples: highly efficient process, fast moving particle). A hyphen rarely follows the letters "ly"—a feature of many adverbs.

- An adjectival phrase, or modifier, before a noun that comprises a number and a unit (of measurement) should only be hyphenated in technical writing when the unit is spelled out (e.g., sixteen-millimeter film, 16-millimeter film, but not 16-mm film).

- Many technical words are formed by adding a suffix (such as anti, de, geo, hypo, post, super, ultra) to a stem word (e.g., antistatic, decalcify, geostationary, hypothermic, postprocessor, supersaturate, ultrasound). These words are generally written without a hyphen; exceptions arise with vowel combinations that would otherwise be difficult to pronounce (e.g., anti-inflammatory, hypo-allergenic, re-engineer, de-ice) or when a consonant is repeated (e.g., shell-like, post-tension). If the stem word is a proper name, spelled with an initial capital letter, then a hyphen is used (e.g., pre-Newtonian, pan-Asian).

- When two related compound phrases are linked together (say with a conjunction), the hyphen is retained after the first word (e.g., pre- and post-test inspection).

- When a prefix or suffix is added to a symbol or numeral, it is hyphenated (e.g., quasi-P, 20-fold, but twentyfold). Similarly, when letters or numbers are used as a prefix or suffix, a hyphen is used (e.g., X-shaped; α-amino; 1,3-butadiene).

- Compound units of measure are hyphenated when spelled out (e.g., newton-meter, foot-pound, passenger-km, kilowatt-hour); however, this does not apply to mile per hour, joule per second, volt per ampere, and so on.

- Compound numbers less than 100 and fractions are traditionally hyphenated when they are spelled out (e.g., sixty-eight, two-thirds).

- Mixed numbers, comprising whole numbers and fractions, are best hyphenated when the numerals are written full-size (e.g., 5¾ is unambiguous, but 5 3/4 is a problem and it would be better to write it as 5-3/4 if small fonts were not available).

Hyphenation: word division at the end of a line

- If possible, avoid dividing words at the end of a line, but in tables and narrow columns, this is often unavoidable.

- Divide words (1) after a prefix (e.g., pre-mature), (2) before a suffix (e.g., assess-ment), or (3) between syllables (e.g., advan-tage).
- Do not split one-syllable words, abbreviations, proper names, dates, or numerals and their associated units.

Hypothesis

A hypothesis is a proposed explanation or premise that may be used as a framework or starting point for an investigation, usually based on limited or inconclusive evidence. If appropriate, the hypothesis should be set out in the **objectives** and, following the presentation of the **results,** should be objectively evaluated.

Be careful in your choice of words—never state that the results *prove* or *disprove* the hypothesis: favorable results will *support* or *uphold* your supposition, while unfavorable results will *contradict* or *refute* the hypothesis.

I

Ibid.

This abbreviation, from the Latin *ibidem*, means *in the same source*. It is traditionally used as a shortened reference in a footnote or endnote (i.e., humanities citation style), where the work cited is the same as that cited in the note immediately preceding it. The abbreviation, which would follow the name of the author(s), eliminates the need for full bibliographical details to be repeated; however, the appropriate page number (or section, figure, table number, etc.) in the original work could be indicated. (The use of footnotes or endnotes for referencing in scientific and technical works is not popular; besides, the abbreviation is regarded as old-fashioned.)

See also **Citing references (basic rules) [footnotes/endnotes]** and **Op. cit.**

Incorrect word usage (confusables)

"Typos" can be brushed aside as little mistakes—minor lapses in concentration, perhaps—but using the wrong word, when something else is intended, hints of ignorance. And to compound the issue, spelling checkers (in word processors) usually do a good job, but grammar checkers seldom pick out an inappropriate word.

There are a relatively small number of words that get mixed up rather frequently in reports (and a few words that should not be used at all as they are ambiguous)—see below.

advice, advise

advice is always a noun (e.g., take the expert's advice); *advise* is always a verb (e.g., to advise the players).

aeroplane, aircraft, airplane

aeroplane is the British English spelling of airplane; an *aircraft* (plural: aircraft) is a heavier-than-air vehicle capable of flight (e.g., airplane, helicopter, glider); an *airplane* is a fixed-wing aircraft.

affect, effect, effects

affect is primarily used as a verb to mean "make a difference to" or "have an effect on" (e.g., education will affect one's earning potential; allergens affect asthma sufferers).

effect, as a verb, means "bring about" a result or change, or cause something to happen (e.g., the sports committee will effect the rule change; growth in sales was effected by the new policy); as a noun, it (1) means a result or consequence brought about by an action or other cause (e.g., the desired effect was realized; Bach's music can have a calming effect), (2) refers to something becoming operational (e.g., it will take effect in two weeks), or (3) indicates impact or success (e.g., morphine had an almost immediate effect).

effects, as a noun, means (1) personal belongings or (2) images, scenery, and sound used in a play/film (e.g., special effects).

allude, elude

to *allude* is to suggest, hint at, or indirectly call attention to; but to *elude* is to (1) evade or escape, or (2) fail to grasp (an idea or fact).

allusion, illusion

an *allusion* is an indirect or passing reference; an *illusion* is a deception, false idea, or belief.

alternate, alternative

alternate, as a verb, means occur in turns (e.g., alternate lectures with tutorials); as an adjective, means "every second" time/event (e.g., alternate weekends); as a noun, means substitute, which, unlike *alternative*, does not imply a "choice" (e.g., following system failure, the aircraft diverted to the alternate airport).

alternative, as an adjective, means other possible (e.g., to evaluate alternative concepts); as a noun, means (1) one of a number of possibilities (e.g., aspartame can be an alternative to sugar) or (2) unconventional (behavior).

amend, emend

both *amend* and *emend* mean improve through correction or adjustment, but *emend* only applies to something written.

and/or, or

X *and/or* Y means X or Y or both, which is not the same as X *or* Y. (*And/or* is sometimes criticized as inelegant, but in technical writing it serves a useful purpose.)

any one, anyone

any one means any single item/person (e.g., removal of any one of the parts will cause the machine to malfunction).

anyone means anybody (e.g., Does anyone want a second-hand book on technical writing?)

any way, anyway

any way means (in) any direction or any manner (e.g., please improve this manuscript in any way you see fit).

anyway means in any case or regardless of any other issue (e.g., she hated me, but I went to see her anyway).

appraise, apprise

appraise means assess or determine (the value or quality of something).

apprise means inform (e.g., she will apprise the President on the state of the economy).

auger, augur

an *auger* is a tool for boring holes; *augur* means an "omen of" or a prediction of some action/event (e.g., recent successes augurs well for the future).

biannual, biennial, biweekly, bimonthly, biyearly

biannual means twice a year and *biennial* means every two years.

biweekly/bimonthly/biyearly can mean either twice a week/month/year or every two weeks/months/years, hence these terms should be avoided.

censor, censure

censor, as a verb, means gag or remove parts (e.g., of a film) in order to suppress the content; as a noun, means a person who censors material (e.g., books, films).

censure means express disapproval (by way of a formal reprimand).

cents, sense

cents are money; *sense* is used in all other contexts (e.g., sense of smell; good sense).

chord, cord

chord means (1) in mathematics, a line across ends of an arc; (2) in aeronautics, width of an aerofoil; (3) in civil engineering, principal member of a truss; or (4) in music, group of notes. In anatomy, it is a variant spelling of cord.

cord means (1) length of rope/string/material/flex or (2) anatomical structure (e.g., spinal cord).

complacent, complaisant

complacent means satisfied (e.g., about one's achievements), smug or unworried.

complaisant means eager to please (others).

complement, compliment

complement, as a verb, means add something as an additional or contrasting element; as a noun, means (1) something additional or contrasting (to enhance or improve) or (2) number required to make up a group/staff/team. In mathematics, it means (1) the amount by which an acute angle is less than 90° or (2) the members of a set that are outside a subset.

compliment, as a verb, means praise or congratulate; as a noun, is an expression of praise or admiration.

composed of, comprises

(is) *composed of* and *comprises* are widely used as synonyms meaning consists of or is made up of (e.g., the set is composed of four CDs; the set comprises four CDs). Whereas it is acceptable to say "the test is comprised of five subtests" or "the test will be comprised of five subtests," it is incorrect to say "the test comprises of five subtests" or "the test comprised of five subtests." (In traditional usage *comprise* means consist of and *compose* means constitute (a whole).)

continual, continuous

continual means frequent or regular (occurrence), but unlike *continuous*, which means without a break or a change (e.g., in character), *continual* implies a series of events with breaks.

credible, credulous

credible describes someone/something who/that is believable or trustworthy (e.g., the case was dropped as the accused had a credible alibi).

credulous describes someone who would too easily believe something (e.g., only the credulous believed the politician's election promises).

criteria, criterion

criteria is the plural of *criterion*, which is the measure or basis by which something or someone is assessed or evaluated.

criticize, critique

criticize means assess the merits and demerits (of something or someone). (It is mainly used in a negative sense, i.e., to find fault.)

critique as a noun, means assessment or criticism; as a verb, means review, assess, or criticize. (In traditional usage, critique would be a noun, and its use as a verb in formal writing is often discouraged.)

data, datum

see **Grammar and style [plurals]**.

defuse, diffuse

defuse means disconnect the fuse (from an explosive device). (It is commonly used in a figurative sense.)

diffuse, as a verb, means disperse, spread, or scatter (e.g., the dye will diffuse in the water); as an adjective, it means spread out.

dependant, dependent

dependant is a noun meaning a person who is reliant on someone else (e.g., his dependants squandered his money).

dependent, as a noun, is an accepted variant spelling of *dependant*; as an adjective, means reliant or contingent (e.g., the outcome will be dependent on the effort).

deprecate, depreciate

deprecate means disapprove; *depreciate* means reduce (in value).

disc, disk

disc is the British English spelling of *disk*.

discreet, discrete

discreet means tactful or avoiding causing offense; *discrete* means separate, distinct, or disconnected.

eminent, immanent, imminent

eminent means famous or admired.

immanent means (1) (a quality) permeating or moving through something, or (2) omnipresent (with respect to God).

imminent means due to happen soon.

enquire, inquire

both words mean ask for information and are used in British English (along with *enquiry* and *inquiry*); while only *inquire* (and *inquiry*) is common in American English. (In traditional usage *inquire* implied a formal request and not a casual question concerning, for example, the weather.)

ensure, insure

to *ensure* is to make certain, but to *insure* is to (1) protect against financial loss or (2) secure payment.

every one, everyone

every one means every single one (e.g., the players were sick, every one of them).

everyone means everybody (e.g., they still won, much to everyone's surprise).

farther, further

when used to mean to, or by, a greater distance (in the literal sense), the words are interchangeable; although *farther* would traditionally be more common (e.g., travel farther); additionally, *further* also means (1) additional (e.g., further tests; further 30 minutes) or (2) advance or progress (e.g., further one's career).

fewer, less

fewer, the comparative form of few, is used with countable things (e.g., fewer people; fewer tests).

less, meaning a smaller amount, is used with (1) mass nouns, which cannot be counted (e.g., less talent; less air), (2) numbers (e.g., less than 50), or (3) expressions of measurement (e.g., less than 220 V).

flammable, inflammable

both words mean "easily set on fire"; rather use *flammable* and *non-flammable.*

flaunt, flout

flaunt means show off or exhibit in an ostentatious way; *flout* means defy or disregard (e.g., a rule or law).

fluid, liquid

fluid denotes matter that flows, including gases, whereas *liquid* denotes fluids but excludes gases.

formally, formerly

formally is an adverb that implies a formal style/mode; *formerly* is an adverb that means in the past (i.e., a former time).

hangar, hanger

a *hangar* is for aircraft; a *hanger* is for clothes.

imply, infer

a writer may *imply* something, which is to strongly suggest something without explicitly stating it.

a reader may *infer* something, which is to deduce something based on reasoning and non-explicit statements or information.

ingenious, ingenuous, disingenuous

ingenious means clever, innovative, or original; *ingenuous* means innocent, naive, candid, or unsuspecting; *disingenuous* means insincere, untruthful, or not candid.

its, it's

its is a possessive pronoun (e.g., a deciduous tree will loose its leaves).

it's is a contraction meaning it is or it has. (Contractions should not be used in formal writing.)

last, past, passed

as an adjective describing a period of time, *last* means most recent (e.g., last year); but *past* refers to a period of time leading up to the point of writing or speaking (e.g., past six years).

passed is the past tense of the verb pass (e.g., it passed through; she passed a remark; it passed the test).

lead, led

lead (pronounced "leed"), as a verb, means to cause to happen or move (e.g., it will lead to; to lead the way); as a noun, means (1) first place (in a contest or list of events), (2) the chief role (e.g., in a play), (3) an electrical wire, (4) an artificial watercourse, or (5) the distance advanced by a screw or propeller in one turn.

lead (pronounced “led”) is a ductile metal (symbol Pb).

led (pronounced “led”) is the past tense of the verb lead (e.g., it led to failure).

licence, license

licence is the British English spelling of the noun license; *license* can be a noun (e.g., pilot’s license) or a verb (to license a procedure).

loose, lose

loose, as an adjective, means (1) not attached/fastened/tied/held together (e.g., a loose button), (2) not fitting tightly (e.g., loose fit), or (3) not exact or poorly defined (e.g., loose schedule); as a verb, means (1) untie, free, or release (e.g., to let loose, to make loose) or (2) discharge (a weapon) (e.g., let loose the arrow).

lose is a verb meaning (1) mislay (e.g., lose a coin), (2) fail to keep/maintain (e.g., lose interest), (3) fail to win (e.g., lose a contest), or (4) reduce (e.g., lose weight).

militate, mitigate

militate (against) means “act as a powerful measure to prevent” (something happening) (e.g., destruction of their habitat will militate against the survival of many rainforest species).

mitigate means alleviate or make less severe/chronic/painful (e.g., conservationists work to mitigate the impacts of human activities on the environment). (Mitigate should not be followed by “against.”)

on to, onto

as prepositions, *on to* and *onto* are used interchangeably to mean “such as to take a position on” (e.g., step on to the platform; walk onto the stage).

on to (and not *onto*) should be used when on is an adverb (e.g., moving on to the next chapter; the bus goes to Milwaukee and then on to Chicago.

practice, practise

practice can be a noun (e.g., in theory and in practice; it gets better with practice) or a verb (e.g., to practice tennis).

practise is the British English spelling of the verb practice.

principal, principle

principal, as an adjective, means main (e.g., principal reason/objective/outcome); as a noun, means (1) the head of a school, organization, et cetera or (2) a sum of money.

principle is always a noun, meaning (1) fundamental source or basis for something (e.g., governing principle), (2) general scientific/mathematical law/theorem, (3) fundamental truth, or (4) in chemistry, an active or characteristic constituent.

shear, sheer

shear, as a verb, means (1) cut (e.g., with scissors or shears) or (2) break off (under structural strain); as a noun, describes a lateral stress or strain (i.e., when layers are shifted relative to each other).

sheer, as an adjective, means (1) perpendicular (or nearly so), (2) very thin (typically fabric), or (3) nothing other than (e.g., sheer concentration); as a verb, means swerve or deviate off course (typically of a boat).

silicon, silicone

silicon is a non-metallic material with semiconducting properties (symbol Si).

silicone is a polymer containing silicon (typically used to produce adhesives, flexible products, lubricants, and sealants).

stationary, stationery

stationary is an adjective meaning not moving; *stationery* is a noun meaning writing materials.

straight, strait, straits

straight, as an adjective, means (1) without deviation (e.g., straight is an arrow), (2) frank or honest (e.g., be straight with me), (3) conventional, or (4) undiluted (e.g., straight whiskey); as a noun, means section of track (e.g., of a racecourse).

strait means tight or restricted; *strait* (or *straits*) means narrow passage of water (e.g., linking two seas); *straits* also expresses a difficult or unpleasant situation (e.g., in dire straits).

systematic, systemic

systematic means methodical or in accordance with a system (e.g., the engineer was systematic in making safe the unexploded bomb).

systemic means universal or in relation to the whole (body or system) (e.g., the infection was systemic, not localized).

that, which

see **Grammar and style [restrictive and non-restrictive clauses]**.

their, there

their means owned by, belonging to or associated with, people or things (e.g., their test).

there is an adverb indicating in, at or to, a place or position (e.g., over there; do not go there).

to, too

to is a preposition expressing (1) a destination (e.g., drive to town), (2) a location (e.g., to the left), (3) a range (e.g., 2000 to 2005), (4) a relationship (e.g., attached to), or (5) a reaction (e.g., to their surprise).

too is an adverb meaning (1) excessive (e.g., too fast) or (2) in addition or as well as (he too measured it).

weather, whether

weather, as a noun, means the state of the atmosphere; as a verb, means to wear away (by exposure to the atmosphere).

whether expresses doubt or choice between alternatives (e.g., whether or not).

who's, whose
who's is a contraction meaning who is or who has. (Contractions should not be used in formal writing.)
whose is (1) used as a determiner to ask questions (e.g., whose book is this?) or (2) a possessive pronoun (e.g., the man whose car broke down is walking).

Internet

There is a view (held by an alarming number of undergraduate students) that the Internet is an unlimited source of information, ready to be downloaded and reformatted into good clean report material. Not only does that constitute **plagiarism**, but it may also be the fastest route to a fail-grade. On the one hand, you can find professional organizations publishing books, journals, scientific articles, and so forth on the Internet, and on the other hand, your search engine can lead you to Joe Bloggs's homepage, which may be full of inaccuracies and "borrowed" material. The rule is simple: before you hit the download key, check carefully who is responsible for the website.

As with any reference, it is important to acknowledge the source and to provide full reference details. An entry in the reference list of simply *The Internet* is ludicrous, considering that there are literally millions of websites. Give the reader a good chance to find the website even if the URL (web address) changes, by providing comprehensive details, as outlined in **Internet reference citation**.

See also **References (non-archival)** and **URL**.

Internet reference citation

In the text of your report, cite Internet documents (e.g., websites and electronic reports, databases, journal papers) in the same way as you would cite print items (see **Citing references (basic rules)**). The reference details (required for the reference list) are no different: there should still be an author or responsible organization, title, publisher, place of publication, date of publication, description of the extent/size (i.e., number of pages) and details on how to obtain the item.

The problem with Internet documents is that (1) some of the reference information may be difficult, or even impossible, to locate; and (2) the URL (web address) or content could change after the website is accessed. It is thus crucial that you provide whatever details you can glean from the website and not just point the readers to the URL.

The order in which the information is presented (and the punctuation) should follow the conventions described in **References (basic rules)** and should be consistently applied.

Internet references: author

Although journal papers, conference proceedings, and books published on the Internet state the name(s) of the author(s), many run-of-the-mill websites do not provide details on authorship. Furthermore, the webmaster or contact person, whose name may be indicated, is frequently not the author. In such cases, the name of the organization serves as both author and publisher. If the name of the author is given, for example, on a personal homepage, then it should be used.

Internet references: title

Books, reports, journal papers, and conference proceedings published on the Internet have clearly identifiable titles, and in the case of journals and proceedings, the name of the journal or conference would also be available. This should be indicated in the reference details.

On the other hand, articles found on websites of individuals or organizations frequently do not have precise titles. In this situation, the name of the individual or organization could be used—for example: *Homepage of Widgets Prosthesis Inc.* It is also acceptable, in the absence of an article title, to construct one by using keywords or a section heading—for example: *Widgets Prosthesis develops innovative design.*

Internet references: revisions/updates

Internet documents use numerals, letters, or revision dates to indicate updates (e.g., Release 7.11; Version B.3; Revised: Jan 18, 2005; Last modified on June 16, 2004). This information should always be indicated.

Internet references: dates

An Internet document could have as many as five different dates, as listed below. If you are using the author-date referencing method, use the publication date for the citation; if this is not available then use the next most appropriate date (working down the list below).

Internet document dates

Date	Notes
When the work was originally published	If available, this should always be indicated in the reference details.
When the work was copyrighted	In the absence of a publication date, the date of copyright can be used.
When the work was placed on the Internet	This date can also be used in the absence of a publication date. If the date is not displayed on the screen, check if it is given in the source code (which can be accessed via the web browser).
When the website was last revised/updated	Revision details, if available, should always be indicated (in addition to the publication date).
When the website was accessed by the researcher	This should always be indicated for websites in which the content may change after initial "publication." However, it is not required for books, reports, journal papers, conference proceedings, and so forth that are published on the Internet with complete reference details.

Internet references: location and extent

Due to the unique structure of web-based documents, volume, chapter, and page numbers are seldom used; but if there is a means to identify the location (i.e., within the larger document) and the extent (i.e., length or size) of the item cited, then this should be indicated.

Internet references: place of publication

For print items, the city in which the publisher resides is sufficient information for the readers to locate the publisher; however, as almost anyone with a computer can become an Internet publisher, an expanded address is usually needed—but it should be kept brief. When a website has been developed for a client, the client's details (not the developer's) should be indicated (as the client is ultimately responsible for the content).

Internet address

The URL is the Internet address (see **URL**). It only specifies a server location, and the constituent elements do not necessarily indicate any hierarchical relationship that can be used for citation purposes. For example, the website www.slezi.hardnox.edu/ could be a mirror site of the SLEZI Corporation hosted by Hardnox University, in which case it would not be a subordinate element of the University's site.

Always verify that the URL that you have indicated can be accessed directly (it is sometimes necessary for viewers to first access a homepage or complete a registration process before they can view other pages). In such cases, you need to outline the process and, if necessary, provide a second URL as the entry point to the website of interest.

Some URLs are incredibly long and would extend onto a second line in the reference list: do not introduce a hyphen or any other symbol to join the two parts; simply break the URL at a suitable location (such as at a slash) and continue on the next line.

Internet references: medium descriptor

Traditionally, a reference work that is not in print format would have a medium descriptor, typically in brackets or parentheses, placed after the reference details—for example: *[videocassette]* or *[proceedings on CD]*. Adhering strictly to this convention requires *[Internet]* to be added as the medium descriptor, and this is occasionally done, but is not necessary as the information is redundant when a URL is indicated.

Introduction

- The introduction starts on a new page. Whereas the **abstract** is punchy, to the point, with no redundant words, the introduction is written in a flowing, easy-to-read style. It places the report in context and focuses the reader's attention on the topics that will follow.
- The introduction is constructed as a story that serves to answer four questions for the reader:
    (1) *What*? Outline what was done and explain the constraints and limitations of the work.
    (2) *Why*? Explain the objectives and the terms of reference for the work.
    (3) *How*? Give a brief overview of the approach adopted.
    (4) *Where?* Give relevant location details, if appropriate.
- It provides background information concerning the tasks undertaken. It should explain the constraints and limitations—both financial and technical—under which the work took place. It puts the study in perspective with respect to the broader scientific or technical picture and describes similar studies that have been conducted.
- If the **background**, **literature review**, or **theory** elements of the report contain substantial amounts of information, then these elements should be placed in their own chapters (which will follow immediately after the introduction); otherwise, they are covered in this chapter.
- The last part of the introduction should prepare the readers for what is coming next (i.e., the main body of the report). This may be placed under an appropriate subheading (called *report structure*, for example). Briefly describe the content and organizational structure of your document—this is particularly important for long or complex

works, such as a PhD thesis. Provide an outline of the content using descriptive terms.

Example:

> The objectives are defined in Chapter 2. In Chapter 3, the relevant theory is described, with supporting details provided in Appendix A. Chapter 4 deals with the experimental procedure and has five main elements, which are ....

- Schematics can be exceptionally useful in providing a roadmap for the readers to understand, for example, the approach adopted or the scope of the test program and its interdependencies. Mention any supplementary items (e.g., a CD containing a computer program) included with the report.
- It is a good idea to write a draft of the introduction early (and revise it later), as there is almost always a rush at the end to get everything finished in time.

ISO standards (for writing)

There appear to be standards for just about everything in engineering! A list of some useful standards for writing reports, from the International Organization for Standardization (ISO), follows:

Some useful ISO standards

Standard number	Title
ISO 31	Specification for quantities, units and symbols
ISO 690:1987	Documentation – Bibliographic references – Content form and structure
ISO 690-2:1997	Information and documentation – Bibliographic references – Part 2: Electronic documents or parts thereof

Some useful ISO standards (*continued*)

ISO 832:1994	Information and documentation – Bibliographic description and references – Rules for abbreviation of bibliographic terms
ISO 2108:1992	Information and documentation – International standard book numbering (ISBN)
ISO 3297:1986	Information and documentation – International standard serial numbering (ISSN)
ISO 8601:1988	Data elements and interchange formats – Information interchange – Representation of dates and times
ISO 14962:1997	Space data and information transfer systems – ASCII encoded English
ISO 15489:2001	Information and documentation – Records management

Italicization

- Italic typeface (*like this*) is used in place of upright roman typeface (like this) to emphasize selected words.

  Example:

  The cylinder exploded prematurely *because* the loads were incorrectly calculated.

- In technical writing it is customary to use italics for
  - Names of individual boats, ships, vehicles, trains, aircraft, spacecraft, and so forth;
  - Foreign-language words—however, words that have been adopted into English (and appear in a standard English dictionary, for example: ad hoc, vice versa) are written in roman typeface;
  - Scientific (Latin) name of plants and animals;
  - Titles and subtitles of published works: books, journals, conference proceedings, newspapers, periodicals, pamphlets, and so forth; or

  - Variables, constants, and certain functions in mathematical expressions (see **Mathematical notation and equations**).

- Italic typeface may also be used for sub-headings and captions (and for examples in books about writing).

K

Keywords

This is a list of approximately four to eight words that can be used to index the work. It is never required for a thesis, sometimes required for a report and almost always required for a journal paper. It is located after the **abstract**.

L

Laboratory report

For many students this is often the first exposure to the drudge of report writing. Laboratory reports are about as personal as the ubiquitous white lab coat; they are the same the world over. Do not rebel. Just give the readers what they expect to see: a standard report structure (after all, you are not writing it for yourself to read).

Typical structure of a laboratory report

Title page
Summary (*or* Abstract)
Table of contents (*or* Contents)
Lists of figures and tables (*if required*)
Nomenclature (*if required*)
1 Introduction
2 Objectives (*or* Aims)
3 Theory (*if required*)
4 Method (*or* Method and Materials *or* Procedure)
  4.1 Apparatus
  4.2 Materials (*if required*)
  4.3 Test specimens (*if required*)
  4.4 Procedure
5 Results
6 Discussion
7 Conclusions (*or* Conclusions and recommendations)
Acknowledgements (*optional*; *it can also be located after the summary*)
References
Appendix (*if required*)

Variation on the typical structure of a laboratory report

If the work comprised a number of separate and distinct experiments, it is often preferable to have a chapter for each experiment, in which the **method**, **results** and **discussion** pertaining to that particular experiment are described. To tie it all together, the closing **discussion** and **conclusions** chapters would deal with *all* the experiments, under appropriate subheadings if necessary.

See also **Structure of reports and theses** and **Error analysis (measurements)**.

Latinisms

- The use of Latin words and phrases—included not for the purpose of communication, but to impress the reader—is a bad idea. If a familiar, everyday equivalent exists, use it. This does not apply to the tens of thousands of common English words that were derived from Latin, but rather to the less well-known Latin words and phrases that people occasionally use (particularly in academic circles).
- The list of Latin words and phrases that is given below is not a compilation of prohibited words. Many of these terms have specific meanings, particularly in a legal context, and the English version may not convey the precise meaning of the Latin term; nonetheless, when it is possible to use an everyday English word, this should be done.

Latinisms

Latin word(s)	Meaning
a fortiori	for a similar, yet even more convincing reason
a priori	literally, from the former (e.g., indicating a deduction by reasoning from prior statements or principles)
ad hoc	established for this purpose only; improvised
ad nauseam	to the point of nausea or disgust; sickening extent
ad valorem	according to the value (i.e., not the amount)
alma mater	school, college or university that one attended

Latinisms (*continued*)

Latin word(s)	Meaning
bona fide	literally, in good faith; genuine; sincere
carpe diem	seize the day (or moment); take the opportunity
ceteris paribus	other things being equal
cum laude	literally, with praise (for a degree award *magna cum laude* is a higher designation of achievement, while *summa cum laude* is the highest level)
de facto	in point of fact; the reality; the actual situation
ergo	therefore; hence; it is concluded
ex officio	by virtue of office; due to position, not merit
ex parte	from only one side (e.g., in legal proceedings); partisan
ex post facto	literally, from a thing done afterwards; retroactive
in absentia	in the absence of; acting in place of the absent person
in loco	at the place
in re	in the matter of; concerning
in situ	in its original or natural place or position
in toto	totally; on the whole; entirely
inter alia / inter alios	among other things / among other people
ipso facto	by that very fact; without other considerations
magnum opus	great work (e.g., literature, art)
mea culpa	my fault
modus operandi	method of operation; method to operate
mutatis mutandis	with the necessary changes made
passim	scattered (e.g., in an index it would indicate that an item is dispersed throughout the work)
per annum	per year; every year; yearly
per capita	literally, according to heads; per person; each person
per diem	per day; every day; daily
per se	by itself; in itself; for its own sake
post hoc [*ergo propter hoc*]	therefore on account of this (the incorrect inference that because one thing follows something else that the first thing caused the second thing to happen)
prima facie	at first sight; from first view; first impression
pro forma	for the sake of appearance or formality
pro rata	in proportion
quid pro quo	something given in exchange for receiving something
re	with regard to; with reference to; in the matter of
sic	so; thus (used to indicate an apparent mistake in a quotation, i.e., something the writer got wrong)

Latinisms (*continued*)

Latin word(s)	Meaning
sine die	without setting a date; indefinitely
sine qua non	literally, without which not; something essential
status quo [*ante*]	the same or current state (of affairs)
stet	let it stand (used to cancel a change, say in editing)
verbatim	word for word; exactly the same
viva voce	by spoken word (e.g., an oral examination)

Notes:

1 A number of these Latin terms are found in everyday English (e.g., ad hoc, alma mater, bona fide, de facto, per annum, per capita, per se, pro rata, sic, status quo, verbatim) and, as they are widely understood, there should be no problem with their usage in a report; however, unnecessary Latinisms (i.e., when an everyday English equivalent exists) should be avoided.

2 Latin words, like other foreign-language words, are usually written in italics; however, words that have been adopted into English are usually written in roman typeface (e.g., status quo, vice versa).

3 See **Abbreviations (common)** for Latin-based abbreviations.

Line spacing

Manuscripts prepared on a typewriter traditionally used double spacing. This antiquated notion persists in some academic quarters, but it looks old-fashioned and wastes paper. A line spacing of between 1.1 and 1.5 is widely accepted and is easier to read. Single spacing is good for tables and footnotes. There is one specific occasion when double spacing is useful and that is when someone is going to edit the manuscript by hand: corrections and comments can easily be inserted between the lines.

List of figures/tables

- It is a good idea to make life easy for the readers—so, if there are a large number of **figures** or **tables** in the report, prepare a *list*. Anticipate that a reader may want to refer to a piece of information, but may not know where to find it. Draw up a list of figures and a separate list of tables, indicating the figure/table numbers, the captions/titles and the corresponding page numbers. The lists should be placed after the **table of contents**.

- If the caption or title is unusually long, use a shortened form, as a concise list will reduce the time it takes a reader to locate a particular entry. References or credits in the caption or title need not be reproduced.
- Lists of figures/tables can be generated automatically—and more importantly, updated with a click of a mouse button—by word-processing software, but it does require planning and the use of a consistent **format style** for the captions/titles.

Lists

A short list with short items can be contained within a sentence (as illustrated in **Punctuation [lists]**), but a long list (or one containing long items) is best arranged vertically (as illustrated below).

Format of vertical lists

- Vertical lists are usually indented; be consistent in the distance that lists are set in from the margin and in the spacing between the list identifier and the text (or list entry).
- List entries can be identified by numbers, letters, or bullets. Use numbers when it is important to describe a sequence of events or to enumerate items—for example, following statements such as: *The following steps were taken*, or *Six actions arise from this decision.* If you need to refer to specific list entries later in the report, say in the **discussion**, then numbers or letters should be used. Bullets are preferable in other situations.
- Numerals (Arabic and Roman) and letters are all used as list identifiers, but Roman numerals are only suitable for short lists.

Examples of list identifiers:

1.	1	1)	(1)	i	i)	(i)	a	a)	(a)
2.	2	2)	(2)	ii	ii)	(ii)	b	b)	(b)
3.	3	3)	(3)	iii	iii)	(iii)	c	c)	(c)

Select a style that will not be confused with other numbering systems adopted in the report (for headings or reference citation, for example). Be consistent in the format of lists—for example: do not use 1), 2), 3) in one chapter and (i), (ii), (iii) in another (unless you deliberately want to make a distinction between the lists).

Punctuation of lists

- A colon is always required after an independent clause (or sentence) that introduces a list—for example:

  The following topics will be discussed in this chapter:
  (a) Boundary conditions
  (b) Numerical model
  (c) Experimental technique

- If the introductory clause has a verb (e.g., evaluates, requires, includes) or a preposition (e.g., from, at, on) that directly refers to (or acts upon) the list items, then a colon is not needed—for example:

  Do not forget to
  1) Check the position of the nearest fire extinguisher;
  2) Extinguish naked flames; and
  3) Shut off the fuel supply.

- There are various acceptable forms of end punctuation for list entries: commas, semicolons, and periods are all used. Many writers punctuate vertical lists with short entries in the same manner as lists contained in sentences (see **Punctuation [lists]**). If one or more of the list entries are full sentences, then a period can be used at the end of each entry.

- Long list entries usually start with a capital letter, but if the list comprises only short phrases or single words, it is acceptable to start each entry with either an upper- or a lowercase letter and to have no punctuation afterwards.

Composition (use of words) in lists

- If you use an introductory clause ahead of a list, such as: *In preparing the samples it is important to*, or *Setting up the instrumentation requires*, then ensure that each list entry makes sense when read in conjunction with the introductory clause.
- Information of a similar content (such as that typically presented in a list of instructions) is best presented using a consistent grammatical format—see **Grammar and style [parallelism]**.

Literature review

- Many undergraduate students submit reports with a literature review section that is nothing more than a list of consulted sources, and perhaps a few words about the content, without realizing that their supervisors have no interest, per se, in knowing what they read in preparation for the work reported on. Their supervisors would, however, be interested in relevant facts and quotations being extracted from the literature and intelligently woven into an essay on the state of the art of the field researched.
- A good literature review will answer for the readers a number of questions about previous work:
    (1) What has been done?
    (2) By whom? Where? When?
    (3) How was it done (i.e., what equipment, materials, and procedures were used)?
    (4) Where are the gaps and boundaries in the knowledge?
    (5) How does it all link together?

The last point is the crux of the issue, as it indicates whether the writer understands what is going on: comments on correlations, contradictions, gaps, and limitations of previous research are required.

- Literature reviews are often written as part of the introductory material of a report or thesis. It is not essential that a separate chapter be included, as the review may be satisfactorily described in the **introduction** or **background**. However, if a thorough investigation *was* conducted—say by someone reading for a PhD—then this becomes an important part of the work and it deserves its own chapter in the thesis.
- Another option is to write a **bibliography** with an appraisal of each item reviewed. In this case, a factual statement about the content and scope of the material covered in each source is all that is needed. If this is secondary to the main thrust of the report, place it after the **references** or in the **appendix**.

Logical conclusions

Both deductive reasoning (which infers the particular from the general) and inductive reasoning (which infers the general from the particular) can be difficult to present in a convincing manner in a report. Readers need to be led along your thought process in a logical and well-constructed argument. Such complex arguments can often be presented as a hierarchy of relatively simple statements (or logical building blocks): with a series of relevant observations placed at the base, upon which reasoned arguments are built, leading to the final, and often most abstract, concept at the top.

For a conclusion to be correct, it is necessary that the premises are correct and that the logic used in the argument is also correct. There are, however, many faults in logic that a writer can be guilty of when deducing conclusions; three illustrations of incorrect logic are given below as examples.

Logic: (1) Overgeneralization/exaggeration

If you observed that there were a disproportionately high number of good left-handed tennis players at your local club, it would be incorrect to deduce that left-handed children are more likely to grow up to be better players than their right-handed counterparts. Clearly, this is false and is an overgeneralization.

Logic: (2) Cause and correlation

- Suppose that a study of the drinking habits of French men indicated that a significantly high number of octogenarians had, as a matter of habit, consumed at least seven glasses of Cognac each week. Is this in itself sufficient evidence to conclude that the drinking of Cognac is responsible for their longevity? Not at all. There could be many social or economic factors responsible for this observation (and one of these factors might also have been responsible for the group acquiring a taste for fine French brandy).
- A *causal* relationship between a condition or an event and an observation is difficult to establish; it is much easier to conclude a *correlation* than it is to conclude that an observation was *caused* by some condition or event.

Logic: (3) Badly constructed argument

- It is possible to express certain logical arguments using the symbols *A*, *B* and *C*:

  Premise 1: *All **A**s are **B**.*
  Premise 2: ***C** is an **A**.*
  Conclusion: ***C** is a **B**.*

- This argument is logically correct and, if both premises are correct, the conclusion will be correct. An example of a trivial but correct argument is:

  Premise 1: *All citrus fruits contain vitamin C.*
  Premise 2: *Grapefruit is citrus.*

Conclusion: *Grapefruit contains vitamin C.*

- ❑ The problem of incorrect logic arises when the argument does not take the precise form given above:

  Premise 1: *All fruits have seeds.*
  Premise 2: *A tomato has seeds.*
  Conclusion: *A tomato is a fruit.*

  This may appear to be correct, but the logic is flawed, as illustrated in the following absurd argument, which has the identical construction:

  Premise 1: *All fish can swim.*
  Premise 2: *A frog can swim.*
  Conclusion: *A frog is a fish.*

M

Mathematical notation and equations

The format and notation for writing mathematical expressions is, by necessity, rigidly defined—even minor departures from the accepted conventions can lead to potentially serious misunderstandings. A list of commonly used mathematical signs and symbols is given below, followed by details on how best to write equations.

Mathematical notation

The following conventions apply to symbols that appear either in passages of text or in equations:

- Italic typeface is used for symbols denoting variables (e.g., x, y, z), algebraic constants (e.g., a, A, b, B, k, K, K_v), coefficients (e.g., drag coefficient, C_D), physical quantities (e.g., gravitational acceleration, g), characteristic numbers (e.g., Euler number, Eu), and so forth, when they appear either in equations or in passages of text. A common mistake is to set symbols in quotation marks when mentioned in the text.

    Example:

    do not write: evaluate "x" and "y"

    rather: evaluate x and y

- Upright (roman) typeface is used for numerals (e.g., $y = 2x$, $y = x^2$), most functions and operators (e.g., $y = \sin x$, $f(x) = \mathrm{d}y/\mathrm{d}x$, $y = \log_a x$), and standard mathematical constants (e.g., π , e, i).
- Bold italic typeface is used for vectors (e.g., $\boldsymbol{u}$, $\boldsymbol{v}$).
- Bold upright (roman) typeface is used for matrices (e.g., $\mathbf{A}$, $\mathbf{B}$).

Commonly used signs and symbols

Symbol	Meaning
Algebraic signs, symbols, and operations	
$+$	plus *or* positive (term)
$-$	minus *or* negative (term)
$\pm$	plus or minus *or* positive or negative (term)
$\times$ *or* $\cdot$	times *or* multiplied by
$\div$ *or* $/$	divided by
$=$	is equal to
$\neq$	is not equal to
$\approx$ *or* $\cong$	is approximately equal to
$\equiv$	is identical to
$>$	is greater than
$>>$	is much greater than
$\geq$	is greater than or equal to
$<$	is less than
$<<$	is much less than
$\leq$	is less than or equal to
$\propto$ *or* $\sim$	is proportional to (*use* $\propto$ *rather than* $\sim$)
∞	infinity
$n!$	n factorial
$\|n\|$	modulus *or* absolute value of n
a^n	a to the power n
$\sqrt{a}$ *or* $a^{1/2}$	square root of a
$\sqrt[n]{a}$ *or* $a^{1/n}$	n^{th} root of a
$\exp x$ *or* e^x	exponential of x
$\log_a x$	logarithm of x to the base a
$\ln x$ *or* $\log_e x$	natural (Napierian) logarithm of x
$\log_{10} x$ *or* $\lg x$ *or* $\log x$	logarithm of x to the base 10. (See note 1.)
$\mathrm{Re}\, z$ *or* $\mathrm{Re}(z)$	real part of z, where z is a complex number

Commonly used signs and symbols (*continued*)

Symbol	Meaning
Im z *or* Im(z)	imaginary part of z, where z is a complex number
Σ	summation of terms
Π	product of terms
Calculus and miscellaneous functions	
Δx	delta x *or* finite change in x
$x \rightarrow a$	x tends to a
$\lim_{x \rightarrow a} y$	limit of y as x approaches a
$\cong$	is asymptotically equal to *or* is approximately equal to
f	function f
$f(x)$	value of the function f at x
f'	derivative function f'
$\frac{\mathrm{d}y}{\mathrm{d}x}$ *or* $\mathrm{d}y/\mathrm{d}x$	derivative of y with respect to x
$\frac{\mathrm{d}^n y}{\mathrm{d}x^n}$ *or* $\mathrm{d}^n y/\mathrm{d}x^n$	n^{th} derivative of y with respect to x
$\partial y/\partial x$	partial derivative of y with respect to x
$\int f(x)\,\mathrm{d}x$	indefinite integral of $f(x)$ with respect to x
$\int_a^b f(x)\,\mathrm{d}x$	definite integral of $f(x)$ with respect to x between the limits a and b
D_x	operator d/dx
$\delta(x)$	Dirac delta function (distribution)
H (x) *or* $\varepsilon(x)$	Heaviside function (unit step function)
Trigonometric and hyperbolic functions	
sin x, cos x, tan x, cosec x *or* csc x, sec x, cot x	trigonometric functions of x (sine, cosine, tangent, cosecant, secant, cotangent)

Commonly used signs and symbols (*continued*)

Symbol	Meaning
arcsin *x* *or* $\sin^{-1} x$, arcos *x* *or* $\cos^{-1} x$, arctan *x* *or* $\tan^{-1} x$	inverse trigonometric functions of *x*. (See note 2.)
sinh *x*, cosh *x*, tanh *x*	hyperbolic functions of *x*
arcsinh *x* *or* $\sinh^{-1} x$, arcosh *x* *or* $\cosh^{-1} x$, arctanh *x* *or* $\tanh^{-1}$	inverse hyperbolic functions of *x*. (See note 2.)
Geometry	
$\perp$	is perpendicular to
//	is parallel to
(x, y)	Cartesian (rectangular) coordinates
(r, θ)	polar coordinates
Sets	
$x \in A$	*x* is an element of, or belongs to, set *A*
$x \notin A$	*x* is not an element of set *A*
$A = B$	set *A* is equal to set *B*
$A \subset B$	set *A* is a (proper) subset of set *B*
$A \subseteq B$	set *A* is a subset of set *B*
$A \cup B$	intersection of set *A* and set *B*
$A \cap B$	union of set *A* and set *B*
Matrices	
$\boldsymbol{AB}$	product of matrices ***A*** and ***B***
$\boldsymbol{A}^{-1}$	inverse of matrix ***A***
$\boldsymbol{A}^{\mathrm{T}}$	transpose of matrix ***A***
det ***A***	determinant of matrix ***A***
tr ***A***	trace of matrix ***A***
Vectors	
$\boldsymbol{u} \cdot \boldsymbol{v}$	scalar (dot) product of vectors ***u*** and ***v***

Commonly used signs and symbols (*continued*)

Symbol	Meaning
$\boldsymbol{u} \times \boldsymbol{v}$	vector (cross) product of vectors $\boldsymbol{u}$ and $\boldsymbol{v}$
∇	nabla (vector differential) operator
$\mathrm{grad}\,\phi$	gradient of ϕ
∇^2 *or* Δ	Laplacian operator
Statistics	
$\bar{x}$	mean value of x
σ	standard deviation (sd)
Q_1	first quartile
Q_3	third quartile
$P(x_i)$	probability that x assumes the value of x_i
P.E.	probable error
χ^2	chi-square distribution
t	Student's t-statistic

Notes:

1. Use $\log_{10} x$ or $\lg x$ rather than $\log x$ to avoid possible confusion with $\log_e x$
2. Use the form $\arcsin x$ rather than $\sin^{-1} x$ to avoid possible confusion with $(\sin x)^{-1}$.

Format of equations

- Symbols denoting physical or algebraic quantities should be defined immediately after the equation in which they are first used and also in the **nomenclature**.

  Example:

  $$V = \pi r^2 h \qquad (3.1)$$

  where V is the volume, r is the radius and h is the height.

Alternatively, the symbols can be listed vertically, which is often easier to read. The units or other relevant information can be provided after the definition.

Example:

$$d = ut + \tfrac{1}{2}at^2 \qquad (3.2)$$

where d = distance (m)

u = initial velocity (m/s)

t = time (s)

a = acceleration (m/s^2).

- The symbols can also be defined in the text immediately preceding the equation, either set in parentheses or marked off by commas.

  Example:

  The volume (V) depends on the radius (r) and the height (h), and is given by …,

  or alternatively:

  The volume, V, depends on the radius, r, and ….

- Number the important equations. The (3.2) used above implies that this is the second equation in Chapter 3. This two-part system can save a lot of time over the alternative, which involves numbering the equations consecutively throughout the report. In the text refer to the equation as either equation (3.2) or Eq. (3.2).

- Avoid using abbreviations in place of conventional symbols for physical quantities or constants. For example, the abbreviation PE, for potential energy, is likely to be unambiguous when used in the text, but in an equation it could be misread as the product of P and E. It would be better to define a symbol for this purpose (such as E_p).

- Whenever possible, omit the multiplication operator (i.e., x); use a raised dot, for example: $y = a{\cdot}b + c$ or write the symbols closed up, for example: $y = ab + c$.

- Use the slash operator (i.e., /) rather than the division symbol (i.e., ÷), for example: $y = ab/c$. Alternatively, write: $y = \frac{ab}{c}$. The slash is preferable for equations within lines of text.

- Equations with multiple terms in the denominators should be written in the form $y = a/(bc)$ or $y = ab^{-1}c^{-1}$ or $y = \frac{a}{bc}$ but never as $y = a/b/c$ or $y = a/b \cdot c$.

- Use parentheses to avoid potentially ambiguous expressions—for example, the equation $y = \sin x + z$ would, in most cases, be interpreted as $y = (\sin x) + z$, but it is possible that someone will read it as $y = \sin\,(x + z)$. By anticipating potential misunderstandings of this type, problems can be avoided.

- Insert a single (thin) space on either side of mathematical operators (i.e., $+$, $-$, $\pm$, $\times$, $\div$, $=$, $\neq$, $\approx$, $\equiv$, $>$, $\geq$, $>>$, $\propto$), but the product of two terms, when written without an operator, does not require a space, for example: $y = ab$, nor does the quotient of two terms require spaces when a slash is used, for example: $y = a/b$. Superscripts and subscripts containing multiple terms are best written in a compact form without spaces, for example: $y = x^{n+1}$.

- Use a small font size for fractions or set them in parentheses—for example, write $y = x/\tfrac{1}{2}\,ab^2$ or better still $y = x/(\tfrac{1}{2}\,ab^2)$ or $y = x/\{(1/2)\,ab^2\}$ but never $y = x/1/2ab^2$.

- Align the = symbols when there is a series of equations with a common left hand side.

Example:

$$
\begin{aligned}
f(x) &= \int \frac{x}{a + bx^2}\, dx \\
&= \int \frac{1}{2(a+bx^2)}\, d\left(x^2\right) \\
&= \frac{1}{2b} \log_e (a + bx^2)
\end{aligned}
$$

- If the right hand side is very long and runs to a second or third line, indent each subsequent line by an incremental amount. It is best to break the equation before an operator not enclosed within brackets or parentheses—preferably at a plus or minus operator—and to move the operator to the next line (see below). If the equation is broken at a point of multiplication or division, then the operator should be written as × or ÷ rather than as a raised dot or a slash.

  Example:

$$
\begin{aligned}
C_{M_G} = - C_L\left\{(h_o - h) + \overline{V}'\left(1 - \frac{d\varepsilon}{d\alpha}\right)\left(\frac{a_1}{a}\right)\left(1 - \frac{a_2\, b_1}{a_1\, b_2}\right)\right\} \\
+ C_{M_o} - \overline{V}'\, \alpha_{T_o}\, a_1\left(1 - \frac{a_2\, b_1}{a_1\, b_2}\right) \\
- \overline{V}'\beta a_3\left(1 - \frac{a_2\, b_3}{a_3\, b_2}\right)
\end{aligned}
$$

Equations in the text

- Mathematical equations appearing in passages of text should make sense when the mathematical operators are read as words, and the sentence should be punctuated accordingly.

Example:

> When $n = 2$, the expression $y = k_1 + k_2 x^n$ provides acceptable results.

This is read as "When n equals 2, the expression …."

- Do not start a sentence with a symbol or number if the previous sentence ended with a symbol or number.

  For example, avoid stating:

  > The exponent n was set equal to 2. k_1 and k_2, the steady-state coefficients, were then evaluated.

  Rather, recast the second sentence:

  > The steady-state coefficients k_1 and k_2 were then evaluated.

- If a punctuation mark placed adjacent to an equation could be mistaken for a symbol in the equation, it should be omitted or the format revised (a comma following an equation can appear to be a prime mark on the denominator).

See also **Characteristic numbers**, **Greek alphabet** and **Units**.

Method

- This chapter is written as a cookbook: you describe the utensils, ingredients, and method for baking the cake. The only difference is that instead of instructing the readers by saying: *Stir into the mixture the yolks of two eggs*, you use the past tense and say: *The yolks of two eggs were stirred into the mixture.*

- The purpose of this chapter is to tell the readers exactly how your study was undertaken. You have to explain to someone—with a similar technical knowledge to your own—how to replicate the experiment. There should be enough information, and no more, for that person to repeat the experiment and get essentially the same results.

- Do not mention people's names without good reason. For example: *Ms. Marjorie Daw, senior laboratory technician, drilled a 10 mm hole in the steel plate using the new drill press.* This sentence wastes the reader's time. Remove Ms. Daw's name and thank her in the **acknowledgements**. All that is required is: *A 10 mm hole was drilled in the steel plate.*

- Only in the case of specialist services (say, work conducted outside of your organization) should you mention the provider—for example: *KrF laser ablation (fluence 1.2 J/cm^2) of the surface was performed by Applied Laser Technologies.* An abridged address of the organization may be given in parentheses or as a footnote.

- If you cite a procedure or technique described in the literature—for example: *The samples were prepared according to the procedure outlined by Shoemaker (1985)*—give the earliest reference that you can find for the process.

Method: typical structure

- **Apparatus** Briefly describe the equipment or hardware used. For sophisticated equipment (i.e., not standard laboratory equipment), be specific and give the brand name and model number of the item, as this may be crucial in replicating the results. In the case of precision measuring equipment, it is necessary to state the accuracy or calibration details. Custom-made fixtures or equipment should be described: full details, including photographs, drawings, or sketches (if necessary), should be placed in the appendix, with sufficient information provided to remake the items.

- **Materials** List the materials used—and be specific: avoid generic descriptions such as *aluminum* if you actually tested high-strength *2024-T3 aluminum alloy*. The company that supplied chemicals can be identified in parentheses when describing the materials—for example: *TEATS (Alchemy Suppliers Inc., Cambridge) was obtained and used without further purification.*
- **Procedure** Present a chronological account of how the experiment was conducted.

MLA (Modern Language Association)

- The Modern Language Association (MLA) produces style guides and handbooks with a humanities slant—these include:
    - *MLA Handbook for Writers of Research Papers*. 7th ed. New York: Modern Language Association of America, 2009.
    - *MLA Style Manual and Guide to Scholarly Publishing*. 3rd ed. New York: Modern Language Association of America, 2008.
- See also **Style manuals**.

N

Nomenclature

- Provide a list of all symbols and abbreviations that appear in the report (excluding common abbreviations used in everyday writing), under an appropriate heading (e.g., *nomenclature*, *list of abbreviations*, or *abbreviations and symbols*). This is located after the **table of contents** and the **list of figures/tables** (if present). Alternatively—and this is less common—it is placed at the start or end of the **appendix** (i.e., where the readers can easily find it).
- There is no prescriptive order for the list, and the following is only a suggestion: the list is alphabetical, with uppercase letters preceding lowercase; Greek letters follow, again with uppercase preceding lowercase; and finally special symbols and subscripts are placed at the bottom of the list. See **Alphabetical arrangement (of lists)**.
- See also **Chemical elements, compounds, and symbols** and **Mathematical notation and equations**.

Numbering system (for chapters and sections)

- The best rule for numbering chapters and sections is KISS (Keep It Simple Stupid).
- Use standard Arabic numerals (i.e., 1, 2, 3, etc.) rather than Roman numerals or Greek letters. Only chapters and sections and subsections within chapters are numbered; the abstract, table of contents, nomenclature, references, and appendices are not chapters, and are generally not numbered.
- Sections within Chapter 1 are numbered 1.1, 1.2, 1.3, etc. and the subsections are designated 1.1.1, 1.1.2, 1.1.3, etc.
- Allocating *numbers* to chapters and *letters* to appendices eliminates any potential mix-up when referring to a specific figure, table, or

section in the report. The appendices are best identified as A, B, C, etc. Sections within Appendix A would be designated A.1, A.2, A.3, and so forth.

- See **Table of contents**, **Headings**, and **Structure of reports and theses** for examples.

Numbering system (for figures and tables)

- To avoid tearing your hair out on the day before the submission deadline for your 100-page research report—when you realize that you omitted a figure or table, and now have to renumber everything—use a two-part numbering system for all long reports. The first part designates the chapter and the second is allocated sequentially within the chapter—for example: *Figure 4.2* (or *Figure 4-2*) will be the second figure in Chapter 4. Note that the first part of the number is always the *chapter* and not a section within a chapter (e.g., do not use *Figure 4.1.3* for the third figure in Section 4.1). Following the same format, the first figure in *Appendix C* is designated *Figure C.1* (or *Figure C-1*).
- If a figure comprises a series of individual elements, identify each part as *(a)*, *(b)*, *(c)*, etc. or *(i)*, *(ii)*, *(iii)*, etc. and have a caption for each one.
- Be consistent throughout the report in regard to the numbering of figures and the format of captions.
- See also **Figures** and **Tables**.

Numbers

Numbers: words versus numerals

The use of words (i.e., one, two, three, etc.) instead of numerals (i.e., 1, 2, 3, etc.) in passages of continuous text (i.e., not in tables, figures and footnotes) is a matter of convention, with no universal agreement existing. Nonetheless, the following rules are widely adopted for scientific and technical writing and may be used as a guideline:

Whole numbers from zero to nine (inclusive) are written as words; all other numbers are written as numerals—with the following exceptions:

(1) Numbers beginning a sentence should be written as words (if this proves tedious, recast the sentence). This also applies to ordinal numbers, which designate position in a series (e.g., first, second).

(2) Values obtained by measurement or calculation appear as numerals (values with units and dimensionless values should be written as numerals).

(3) When two numbers are adjacent in a sentence, spell out one of them (e.g., twelve 10 mm bolts, twenty-one 50 g samples).

(4) Numerals are used when a number less than 10 appears in a range or list with higher numbers (e.g., write *specimens 4, 11, and 17* rather than *specimens four, 11, and 17*).

(5) Numerals are used in reference to numbered elements of a document (e.g., chapter 5, page 9) or coordinates in tables (e.g., column 2, rows 1–4).

Numbers: style and format

- Hyphenate compound numbers when they are spelled out (e.g., fifty-two measurements were taken).
- Most authorities recommend putting a space between a numeral and its unit (e.g., 3 m/s) rather than writing them "closed up" (e.g., 3m/s)— this is, nevertheless, a matter of style preference, but the adopted format should be consistently applied throughout the report. There is an exception: it is customary not to put a space between a numeral and a superscript-type unit symbol (e.g., 90°, 30" 5'); however, when used as a measure of temperature the degree symbol

can be preceded by a space or it can be written "closed up" (e.g., 20 °C, 20°C). See also **Units**.

- In reports written for an international readership, it is better to not use commas as separators in large numbers (to mark thousands, millions, etc.). The comma is used as a "decimal point" in many countries and, for example, the number 96,415 could easily be misunderstood. Blank spaces, which are used instead of commas, can be inserted both before and after the decimal point (e.g., 73,296.4158 can be written as 73 296.415 8).

- Avoid using the words *million*, *billion*, *trillion*, *quadrillion*, et cetera with engineering units (e.g., don't write *3.2 million km*); instead, use scientific notation (e.g., 3.2×10^6 km). In old British texts a billion denoted 10^{12}, but today the American definition of 10^9 is universally used.

- If the numeral *zero* could be mixed up with the letter *O*, or the numeral *one* with the letters *l* or *I*, spell out the number.

- In the past, ordinal numbers were always spelled out in formal writing (e.g., written as *twenty-seventh* and not *27th*), but this is cumbersome, and most technical and scientific editors have discarded the convention. However, in line with the custom of spelling out numbers less than 10 (when they appear in passages of continuous text) the ordinal numbers *first, second, third ... ninth* are spelled out.

- Use *ca.* or the tilde symbol (i.e., ~) to indicate an approximate value if you do not wish to write out the word *approximately* (e.g., the vehicle will have a mass of ca. 450 kg; the time interval was ~ 70 μs). Do not use the symbol ± for this purpose as it literally means "plus or minus" and indicates either a tolerance to a measurement or a number that could be positive or negative.

Numbers: fractions and decimals

- Fractions written out in the text (say at the start of a sentence) should be hyphenated (e.g., three-eighths). If this appears clumsy, recast the sentence and use numerals in fraction format (e.g., ⅜) or decimal format (e.g., 0.375).

- Use a reduced font size for fractions, if possible (e.g., write ¾ instead of 3/4).
- A mixed number, comprising a whole number and a fraction (e.g., 4⅓), is best written without a space between the two elements; however, if the numerals are written full-size, then it is better to use a hyphen (e.g., 5-3/8).
- Do not mix fractions and decimals when making a comparison or defining a range (e.g., do not write *from 4½ to 5.35*).
- Use an initial zero for decimal numbers less than one (e.g., write *0.425 mm*, never *.425 mm*).
- Align the decimal points in columns of numbers.

Numbers: percentages

- Use the symbol % for percentage following numerals, rather than the word *percent* (e.g., 34.5% of the data points). There is no space between the numeral and the % symbol.
- Do not use the informal abbreviation *% age*.
- A *percentage point* change does not have the same meaning as a *percent* change. If a value (representing a rate, for example) increased from 5% to 10%, this would mean an increase of 5 percentage points, but it would be misleading to say that the value increased by 5% as the value doubled (which is a 100% increase).
- If a value has increased by 200%, it has trebled. It is not possible for something to reduce, or to devalue, by more than 100%.

Numbers: ranges

- Indicating a range of numbers using the words *between* and *and* (e.g., between 10 and 17) can be ambiguous (are the limits included?). A better option is to use the words *from* and *through* to connect the numbers (e.g., from 10 through 17); alternatively, be explicit and add *inclusive* or *exclusive* after the range.

- An en dash (which is longer than a hyphen) can also be used to describe an inclusive range of numbers (e.g., pages 199–202, 150–180 Hz, 220–240 V). Units are not normally repeated when a dash is used, but it is popular to repeat the percent symbol (e.g., 75%–80%). As the dash is read as *to*, it is incorrect to write *between 100–200*, for example.
- When using the word *to* to describe a range, write the units twice (e.g., from 500 ft to 1500 ft). Be consistent with units (e.g., avoid *95 cm–1.05 m*, rather write *0.95–1.05 m* or *0.95 m to 1.05 m*).
- Do not omit digits from the second number in a range if there is any chance of misunderstanding (e.g., write *years 1896–1903*, not *years 1896–03*).

See also **Characteristic numbers**, **Abbreviations [date and time]**, **Significant figures**, and **Units**.

O

Objectives

- This chapter may be called *objectives* or *aims* or even *aims and objectives*. Aims are sometimes described as long-term objectives, but in most situations the terms are used synonymously (in which case the heading *aims and objectives* is partially redundant). Less popular headings for this chapter are *purpose* and *scope*, which do not have the same meaning as *objectives*.

- The chapter should contain a series of unambiguous statements of what you set out to do (e.g., to demonstrate, investigate, calculate, or measure something). Inform the readers why the work was undertaken—for example, this could be to prove or disprove something, to measure certain parameters experimentally or to design, build, and test something. These are the goals that were set and they must be concisely and accurately stated.

- The objectives do not necessarily have to be listed in the chronological sequence in which the work was performed, but should rather be structured in a logical sequence that will make sense to someone who has no knowledge about the work that you undertook. Primary objectives are usually stated before secondary objectives.

- Refrain from describing how the work was conducted and do not present results or comment on whether the goals were achieved: this comes in later chapters.

- When the report is nearly finished, compare the *objectives* to the *conclusions*; there should be a synergy between what you set out to do and what you concluded from the work. Revise the two chapters accordingly.

Op. cit.

The Latin abbreviation *op. cit.* (from *opera citato*) is used in academic writing to mean *in the work already cited.* It is traditionally placed in a footnote or endnote (i.e., humanities citation style) to make reference to a previous footnote, but not the one immediately preceding it. The abbreviation follows the author's name and eliminates the need for full bibliographical details to be provided each time the work is cited; however, the appropriate page or section in the original work could be indicated following the abbreviation. (The use of footnotes or endnotes for referencing in scientific and technical works is not popular; besides, the abbreviation is regarded as old-fashioned.)

See **Citing references (basic rules)** and **Ibid**.

P

Page numbers (pagination)

- All pages should be numbered, including pages that incorporate figures or tables that take up the whole page. This also applies to blank pages (included, for example, to get the pagination correct for a report printed on both sides of the pages). If the report has more than one volume, the pagination should indicate the volume number.

- It is common for the *introduction* to start on page 1, with all preceding pages given lowercase Roman numerals (i.e., i, ii, iii, etc.). Odd page numbers should be on the right-hand side of reports printed in "book" format (i.e., printed double sided).

- The *appendices* are usually paginated separately from the main text and from each other. The pages can be numbered A1, A2, etc., for Appendix A, for example. The pages can also be numbered A.1, A.2, etc. (with a dot) or A-1, A-2, etc. (with a hyphen); however, if the sections within the appendix are designated as A.1, A.2, etc. and the figures as A.1, A.2, etc. (or A-1, A-2, etc.), then it is better to write the page numbers in a unique format (i.e., without a dot or hyphen), as initially suggested. (This distinction becomes important if you make changes using the *find and replace* word-processing tool.)

- Some organizations like to indicate the total number of pages on the title pages (or front covers) of their reports and also on every page, usually in the format *page # of #*. This is part of **revision control**, a document management process that ensures that users have complete and up-to-date copies of reports.

- The traditional *thesis* format has the page number located centrally at the top or bottom of the page, but there are no fixed rules about this: it is a matter of personal choice.

- Getting the page numbers correct for a large report with many contributors can be tricky—as a last resort, write the numbers in by hand, but you should never leave them out.

Paragraphs

- Bearing in mind that long paragraphs can be a cure for insomnia, break up the text into manageable parts. Each paragraph should deal with one basic theme. Avoid one-sentence paragraphs, if possible, and random paragraph breaks that divide material that naturally belongs together.

- Identify the start of each paragraph by indenting the text or by inserting a line space, but do not switch arbitrarily between one method and the other. Some authors prefer not to indent the first paragraph beneath a section or chapter heading (this is a matter of style preference), but it is important that the adopted format be consistently applied throughout the report.

- Typical paragraph structure:
  - *Opening statement*: which may introduce the topic and provide general information.
  - *Main paragraph body*: which expands on the topic, providing technical details, results, examples, data, comparisons, arguments, and so forth.
  - *Closing statement*: which may present conclusions, revisit the main points of the paragraph, and/or lead into the topic of the next paragraph.

Paraphrasing

To paraphrase is to reword a passage without changing its meaning. A necessary part of almost every academic study is a review of what has been done before: as background to establish the context of the present study, or to compare or contrast methodologies, results, conclusions, and so forth. To describe someone else's work inevitably means that you have to read their publications and summarize the relevant information—but paraphrasing is difficult to do well, as it invariably forces your thoughts down the path taken by the author. It is often better to study the publication in detail—developing a good understanding of what was done—and to write from a "blank sheet," than to work line by line through the source material trying to express each sentence in your own words. Sentences or phrases that are not reworded must be written in quotation marks (see **Quotations**).

Note that paraphrasing provides no immunity from challenges of **plagiarism**: an idea that has been previously described in another work must be credited to the originator, even if you have reworded it.

Passive voice

What is the passive voice?

If the grammar checker (in your word processor) complains about your use of the *passive voice*, what does it mean? Well, it concerns the order of the words in the sentence. The passive voice has the *object* of the sentence—that is, the person or thing to which the action is directed—at the beginning. This is not the case for the *active voice*.

For example:

(1) Passive voice: *The hen was caught by the fox.*
Order: object (*hen*), verb (*caught*), subject (*fox*)

(2) Active voice: *The fox caught the hen.*
Order: subject (*fox*), verb (*caught*), object (*hen*)

Is it incorrect to use the passive voice?

No, it is not incorrect. The grammar checker picks this out because the active voice is more concise, more direct, and easier to read. For these reasons, it is better to write in the active voice.

However, technical reports are impersonal and, as part of the process of excluding the personality of the writer, it has been customary not to use the first-person pronouns: *I* and *we*. (Exceptions occur in the **acknowledgements** and **dedication**, which by their nature *are* personal, and statements such as *I wish to thank ...* are often used.) When authors describe an action in a technical report without the benefit of these pronouns, they frequently resort to the passive voice. For example, the statement in the active voice: *I calculated the mean value*, becomes: *The mean value was calculated [by me].* The *by me* is understood and omitted.

When is the passive voice a problem?

The active voice (sentence 1 below) can be written in the passive voice to avoid using a first-person pronoun (sentence 2). This is perfectly acceptable; the problem arises when the passive form becomes distorted and cumbersome (sentence 3). In this case the sentence is based on an abstract noun (cutting) plus a vague verb (undertaken) rather than a vigorous verb (cut).

Example	Voice	Comment
(1) *I used a diamond saw to cut the specimens to size.*	Active voice	No good for report writing (it uses a first-person pronoun)
(2) *The specimens were cut to size using a diamond saw.*	Passive voice	Perfectly acceptable
(3) *Cutting of the specimens to size was undertaken using a diamond saw.*	"Distorted" passive voice	Cumbersome

Watch out for the words *undertaken*, *performed*, *carried out*, *accomplished*, and *achieved*, which are all indicators of the use of a cumbersome sentence structure. In most cases it will be possible to rewrite the sentences in a simpler, more direct manner.

Examples:

(1) "Distorted" passive voice: *Daily testing of the revised computer program was performed using actual experimental data.*

Passive voice: *The revised computer program was tested each day using actual experimental data.*

Active voice: *The author tested the revised computer program each day using actual experimental data.*

(2) "Distorted" passive voice: *Removal of the oxide layer was carried out immediately prior to bonding using grit-blasting.*

Passive voice: *The oxide layer was removed using grit-blasting immediately prior to bonding.*

Active voice: *The author used grit-blasting to remove the oxide layer immediately prior to bonding.*

If you cannot think of a way of phrasing a sentence in the passive voice, refer to yourself as *the author*, as illustrated in the two previous examples. Just make sure that you have not mentioned another author immediately prior to this statement, which would make the sentence ambiguous.

Be careful with phrases such as: *It is believed that ....* If this is in the context of a discussion concerning experimental results, it may not be clear if this is a statement of widely held belief or your personal opinion based on the results obtained. To avoid the problem, state: *The author was of the opinion that ....* Alternatively an exception to the rule about not using first-person pronouns should be made; after all, there is no point in constructing an ambiguous sentence solely to comply with convention—it is always better to write something that will be clearly understood.

Photographs

- Including photographs of test equipment, specimens, laboratory apparatus, and so forth in a report is a relatively easy thing to do, and it can be an effective means to convey valuable information to your readers. However, there are a number of conventions and pitfalls that may not be immediately apparent:
  - Treat photographs as **figures**: give each one a figure number and a caption.
  - Crop unrelated details from images (badly taken photographs may contain irrelevant details, which will confuse readers).

- Do not use color if the report is to be copied (on a black and white copier) for further distribution.
- Place a ruler (scale rule) adjacent to test specimens so that readers can appreciate the size of the items.
- Ensure that the scale and not just the magnification is shown on photomicrographs and electron micrographs (where the image size is important), as the image may be altered in reproduction.
- Obtain the permission of the **copyright** holder if you wish to use their photographs (using images scanned from printed works or downloaded from the Internet may be unlawful).

❑ Consider that a reader may have never seen the subject of the photograph and that it may be difficult to interpret. Line drawings—which exclude needless detail—are usually easier to understand than photographs.

Plagiarism

Plagiarism is theft. It is the result of including in your report a figure, data, an idea, or a piece of text from another source and failing to acknowledge that fact. By not referencing the original work, you imply that the "borrowed" material is your own, and that is wrong. Avoid the problem by meticulously referencing all elements in the report that are not your own work, including paraphrases. **Paraphrasing** can get around the issue of copyright (which resides in the form of words used), but it does not provide a cover for stealing an idea, which is plagiarism.

See also **Citing references (basic rules)**, **References (basic rules)**, and **Fair-use doctrine**.

Preface

A preface is an introduction to a document that can be used to explain the motivation, background, and context of the work undertaken, and also to pay tribute to those involved in the project. Previous editions of the work, contributions of collaborating authors, and new material in the current edition can be described under this heading.

A preface is more commonly found in published books than technical or scientific reports or theses, where expressions of gratitude are placed in the **acknowledgements**, and the motivation and background are explained in the **introduction**.

Progress report

The writing of progress reports, for submission to funding agencies or external organizations, is frequently viewed by engineers as an irritating distraction from the real work; as a consequence, the reports can be poorly structured with excessive or inappropriate technical detail (possibly pieced together from other technical reports or presentations) and with limited information concerning the project finances and schedule. Furthermore, the bigger picture (say, concerning the implementation of the results of the study) is frequently missed out—not necessarily because the engineers and scientists do not see it, just that they fail to describe it appropriately.

People working for an agency or external organization that funds work under contract want to know how their money is being spent and they want reassurance that the targets (concerning performance, schedule, and cost, for example) will be met.

Strategy for writing a progress report

From the outset, you should

(1) Focus on issues that are important to the funding agency (and not what interests you);

(2) Describe the work performed, milestones reached, delays encountered, deliverables submitted, and resources consumed, all within the context of the contract (or agreed work program); and above all

(3) Ensure that the report can be understood by non-technical people without "dumbing down" the content.

Typical structure of a progress report

Title page
Summary (*or* Executive summary)
Table of contents (*or* Contents)
Lists of figures and tables (*if required*)
Glossary and nomenclature (*if required*)
1 Introduction
2 Objectives (*or* Purpose *or* Scope *or* Terms of reference)
3 Project plan (*or* Schedule *or* Time schedule *or* Workplan *or* Work program)
4 Finances (*or* Costing *or* Cost summary *or* Expenses *or* Resources)
5 Technical progress (*or* Work completed to date *or* Results *or* Achievements)
  5.1 Work package 1 (*individual work elements are discussed under appropriate headings*)
  5.2 Work package 2 (*ditto*)
  5.3 *Et cetera*
6 Discussion
7 Conclusions
8 Future work (*or* Revised project plan *or* Revised schedule)
References (*if required*)
Appendix (*if required*)

See also **Structure of reports and theses**.

Proofreading

- Always proofread the final draft of your report on paper rather than on the computer screen, as you will definitely pick up more mistakes that way. It also gives you a better perspective of what the final product will look like. Ask a family member or friend to read the report. Often someone with no knowledge of the subject does the best job: they are less likely to skip over words because they will not be familiar with the subject matter.

- Force yourself to read aloud, checking the text—line by line (sliding a ruler down the page, if necessary)—for errors in spelling, punctuation, grammar, and so on. A novel approach is to use text-reading software. Listening to a computerized voice reading your report can be amusing, but more importantly the computer will read what you *actually* wrote and not what you *intended* to write.
- Some of the easiest things to mess up are
    - Figure and table numbers and their captions and titles;
    - Citations and the reference list;
    - Cross-references (to other parts of the report);
    - Chapter and section headings;
    - Abbreviations and acronyms;
    - Numerical results (e.g., significant figures and units); and
    - Pagination and table of contents.

Strategy for proofreading

- Flip through the report just looking at the **figures**; check to see if they are sequentially numbered, that all have captions (with the same font size and style) and that an appropriate reference has been made to each one in the text. And then do the same thing with the **tables**.
- Check that all cited **references** are included in the reference list; identify them on the list as you go through the report, and then remove all non-cited references.
- Check the format of **headings** for consistency, and make sure that you do not have a heading at the bottom of a page with no text.
- Ensure that all **abbreviated terms** are defined in the **nomenclature** and, again, in the text at their first mention (use the word processor's search tool for this).

- Scan the report for numerical values and check the format of each one—ensure that an appropriate number of **significant figures** have been used and that the **units** have not been omitted.
- Check that all pages are numbered correctly and that the **table of contents** and lists of figures and tables (if included) are correct and complete.
- Finally, when you have finished editing, sit down and read the whole report again. Last-minute changes have a nasty tendency to introduce spelling mistakes, or to produce inconsistencies in style (prose), referencing or numbering.

Punctuation

The use of punctuation marks depends on the writer's intention: it could be to

(1) Define the nature of a sentence (terminal punctuation), using a *period*, *exclamation point*, *question mark*, or *ellipsis marks*.

(2) Show the relationship between ideas in a sentence, using a *colon*, *dashes*, *parentheses*, and/or *commas*.

(3) Join two sentences, using a *colon*, *semicolon*, or *comma*.

(4) List items within a sentence, using a *colon*, *comma*, or *semi-colon*.

(5) Show a possessive relationship, using an *apostrophe*.

(6) Quote something, using *quotation marks*.

(7) Clarify word usage, using a *hyphen* or *quotation marks*.

(8) Indicate a range or link words, using a *dash* or *slash*.

These topics are covered in this section. For a more comprehensive treatment, consult the publications listed under **Style manuals**. A good starting point, however, is the *Elements of Style* (W. Strunk Jr. and E. B. White, Allyn and Bacon, 4th ed., 2000), the classic "little book" of style, or the *Oxford Style Manual* (R. Ritter, Oxford University Press, 2003). See also **Abbreviations (common)**, **Glossary** (which includes an explanation of grammatical terms), and **Grammar and style**.

1 Define the nature of a sentence (terminal punctuation)

1.1 To identify a statement:

end sentence with a *period.*

1.2 To identify an exclamation:

end sentence with an *exclamation point.*

The exclamation point is used to draw attention to a command, a wish, an assertion, a witty remark, or an obscenity; but it has no place in technical reports.

Example:

The data analysis took 17 hours (phew!).

1.3 To identify a question:

end sentence with a *question mark.*

An objective of a study can be expressed as a question; direct questions are usually set off by commas (e.g. 1 below). Usually it is easier to report the question indirectly, in which case a question mark is not used (e.g. 2). Generally speaking, rhetorical questions—asked not to elicit information, but to make a statement—should not be used in reports (e.g. 3).

Examples:

(1) *The question was asked by the research team, why did the engine attachment fail?*

(2) *The question that the research team asked was why the engine attachment failed.*

Example of a rhetorical question:

(3) *What does this tell us?*

1.4 To indicate an incomplete sentence:

use *ellipsis marks* (i.e., three periods).

An ellipsis can be used by a writer to say "left to your imagination," but it should never be used in this way in a report (e.g. 1 below). It may, however, be used to continue a well-known series (e.g. 2); although, the abbreviation *etc.* would be more common in this role. When an ellipsis ends a sentence, a period may be added after the ellipsis (i.e., four dots without a space).

Examples:

(1) *There was nothing more that could be done: the mains power had failed, there were no torches in the building, the backup generator was broken ....*

(2) *It was convenient to use the radio communication alphabet names: alpha, bravo, charlie ....*

2 Show the relationship between ideas in a sentence

2.1 To introduce an explanation or elaboration:

use a *colon*.

A colon follows what is a complete sentence and introduces something that explains or elaborates on the main idea of the preceding part (e.g. 1 and 2 below). The second clause can be a quotation (e.g. 3). Material introduced by a colon should start with a capital letter if it consists of more than one sentence (e.g. 4).

Examples:

(1) *There were only three things wrong with Murphy's experiment: the apparatus, the materials, and the methodology.*

(2) *His conclusions contravened Newton's second law of motion: acceleration is proportional to the resultant force.*

(3) *The client's assessment was emphatic: "His work has no value, which is the reason why payment is being withheld."*

(4) *The manager faced a dilemma: If she terminated Murphy's contract, she would have to move someone else onto the*

project. On the other hand, by not terminating the contract, his unacceptable behavior would be condoned.

2.2 To introduce an explanation or illustration that runs up to the end of the sentence:

use a *dash* (i.e., an *en dash* or an *em dash*) .

An explanation or illustration can be introduced by either an *en dash,* which requires a space on either side (e.g. 1 below) or an *em dash,* which is usually set without spaces (e.g. 2). As a colon can also be used in this role, a sentence with a dash can usually be rewritten with a colon (e.g. 3). Two dashes in a sentence usually denote a parenthesis (see section 2.3); do not use three or more dashes (e.g. 4).

Examples:

(1) *Dashes come in two sizes – there is a short one called an en dash (–) and a long one called an em dash (—).*

(2) *Often there is more than one way of doing something—a colon could be used here with a slightly rephrased sentence.*

(3) *Often there is more than one way of doing something: here, a dash could be used with a slightly rephrased sentence.*

Example of incorrect usage:

(4) *The over-use of the dash—as an all-purpose punctuation mark—as illustrated in this badly constructed sentence—suggests that the writer is lazy in bringing together his or her ideas—and this makes for poor writing.*

2.3 To insert a word, phrase, or clause (i.e., a parenthesis):

use one of the following three alternatives:

parentheses, *dashes*, or *a pair of commas*.

A parenthesis (an insert into a sentence) interrupts the main flow of the sentence to explain or elaborate on something (e.g. 1–8 below). It can be omitted, leaving a less informative, but grammatically complete, sentence. Use either *em dashes* (—) or *en dashes* (–) for this purpose, not both types (e.g. 4–6). To avoid confusing readers when commas are used, keep parentheses short (e.g. 7 and 8).

Examples (parentheses):

(1) *Parentheses (called round brackets in British English) are probably the easiest way to mark a parenthetical clause.*

(2) *The punctuation mark that follows the closing parenthesis (if one were needed) would be the same as that in the absence of the parenthetical expression.*

(3) *A complete sentence in parentheses (when enclosed in another sentence) should never end with a period; however, an exclamation point or question mark can appear inside parentheses.*

Examples (dashes):

(4) *Some style manuals – and writers – prefer the shorter en dash to the longer em dash.*

(5) *Other writers prefer the longer em dash—which can be set with or without spaces on either side—to the shorter en dash, as it more clearly identifies the parenthetical text.*

(6) *Dashes can be used to boldly mark out parenthetical expressions—facilitating the construction of complex sentences, where the parenthesis is itself punctuated by commas—but be careful, their over-use can suggest carelessness.*

Examples (commas):

(7) *A short parenthetical clause set in commas, one of the many uses of commas, does not interrupt the flow of the sentence to the same extent as a clause set in parentheses or dashes.*

(8) *Commas are, however, the only way of marking a parenthetical adverb (e.g., however, nevertheless, moreover) in mid-sentence.*

2.4 To insert a non-restrictive word, phrase, or clause:
in general, use a *pair of commas*.

A non-restrictive word, phrase, or clause—which provides additional information, but does not define or identify the noun preceding it—is parenthetic (explained in section 2.3 above) and should be marked accordingly (e.g. 1–6 below). A sentence with a restrictive clause cannot be rewritten as two sentences without a change of meaning (unlike a sentence with a non-restrictive clause) and should not be marked by commas (e.g. 7 and 8). See also **Grammar and style [restrictive and non-restrictive clauses]**.

Examples (non-restrictive):

(1) *The epicenter of the shock, where the greatest damage occurred, was 40 km from the capital.*

(2) *An after-shock, which measured 4.8 on the Richter scale, was felt in the capital.*

(3) *The mayor, who returned from his holiday, described the devastation.*

(4) *Andre Du Prez, the renowned photojournalist, was killed.*

(5) *The president, Miguel Ramón, declared an emergency.*

(6) *A tsunami, or seismic sea wave, is rare in this part of the world.*

Examples (restrictive):

(7) *The bridge that was damaged will take four years to re-build.*

(8) *The people who live in the area are facing a bleak winter.*

2.5 To separate subsidiary or introductory information from the main sentence:

use a *comma*.

A comma follows a subsidiary or introductory clause at the start of a sentence (e.g. 1 and 2 below). Similarly, if the first word is an adverb (such as *already*, *however*, *moreover*, *nevertheless*, *therefore*), which can be removed to leave a grammatically complete sentence, a comma is required (e.g. 3).

Examples:

(1) *Even before the final whistle sounded, the supporters were celebrating by singing and waving flags.*

(2) *In fact, the fireworks started after the third goal was scored.*

(3) *Already, the fans are talking about the cup final.*

3 Join two sentences

3.1 To join two sentences where the second explains, elaborates, or contrasts the first:

use a *colon.*

A colon can introduce an explanation or elaboration (as described in section 2.1 above), but it can also be used to join opposing or contrasting statements, usually for dramatic effect.

Example:

The screw would not budge: he was turning it the wrong way.

3.2 To join two sentences with linked ideas without a conjunction:

use a *semicolon.*

A semicolon is used to join two sentences with linked ideas (frequently related by cause and consequence) without a conjunction.

Examples:

(1) *The engine started losing oil. There was little chance of reaching the destination airport.*

(2) *The engine started losing oil; there was little chance of reaching the destination airport.*

3.3 To join two sentences with an adverb:

use a *semicolon.*

A semicolon is required when joining two sentences or clauses with an adverb (such as *besides*, *consequently*, *hence*, *however*, *nevertheless*, *then*, *therefore*, *thus*). A comma often follows the adverb.

Example:

The pilot was forced to divert to a disused airfield; however, the runway was in good condition, and the plane landed safely.

3.4 To join two sentences with a conjunction:

use a *comma.*

Use a comma before a conjunction (such as *but*, *and*, *or*, *nor*, *yet*, *so*, *for*) joining two sentences or clauses (e.g. 1 below). However, when the conjunction is *and* and the subject of the second clause is the same as the first (and the subject is not repeated or replaced by a pronoun), then the comma is best omitted (e.g. 2).

Examples:

(1) *The passengers disembarked, but the pilot stayed onboard and contacted the ground crew.*

(2) *The ground crew arrived after several hours and appeared to understand the problem even before the engine cowling was removed.*

4 List items within a sentence

A long list or one containing long items is best arranged vertically (see **Lists**), but a short list with short items can be contained within a sentence (as illustrated below).

4.1 To introduce a list:

in general, use a *colon.*

A colon is needed if an independent clause—which, by definition, can stand alone as a complete sentence—introduces a list (e.g. 1 below). This means that an introductory clause that cannot stand alone does not need a colon: such clauses frequently contain a verb (e.g., *require*, *consider*) or a preposition (e.g., *from*, *on*) that refers to, or acts upon, the list items (e.g. 2 and 3). A list introduced by the word *following* (or *as follows*) is always preceded by a colon (e.g. 4).

Never use a colon and a dash to introduce a list (the combined mark " :– " is not standard punctuation).

Examples:

(1) *The operation requires four people: a cook, two waiters, and a manager.*

(2) *The operation requires [no colon needed] a cook, two waiters, and a manager.*

(3) *The startup finance comes from [no colon needed] personal savings, a bank loan, and the sale of two vehicles.*

(4) *The following were discussed: preparing the food, serving the customers, and taking their money.*

4.2 To separate items in a list:

in general, use a *comma.*

In a series of three or more elements (i.e., words or phrases), a comma is used to separate elements (e.g. 1 below). Many authorities advocate putting a comma after the second-last item in the list, that is, before the *and* or *or* (e.g. 2 and 3). This is, however, a matter of style preference (it is called a serial or Harvard or Oxford comma). Commas are not needed when conjunctions are used to join all the elements in a series (e.g. 4).

Examples:

(1) *They found a small cafe, sat down, ordered drinks, and read the menu.*

(2) *The salad had cheese, croutons, and sunflower seeds.*

(3) *There was a choice of tea, coffee, or fruit juice.*

(4) *He ordered tea and coffee and fruit juice.*

4.3 To separate adjectives:

in general, use a *comma.*

In a list of adjectives, a comma is used where an *and* would be appropriate (e.g. 1 and 2 below), which is not the case in example (3).

Examples:

(1) *The drink was light, fruity and non-alcoholic.*

(2) *The light, fruity, non-alcoholic drink was consumed.*

Example of incorrect usage:

(3) *It was red, grape juice.*

4.4 To separate list items that contain a comma:

in general, use a *semicolon.*

Semicolons are needed if any item in a series contains a comma (e.g. 1 below). Numbers can be used to clarify complex lists—in which case either commas or semicolons can be used to separate the items (e.g. 2).

Examples:

(1) *The lesson included cooking the steak, eggs, and chips; adding salt, pepper, and sauce; tasting the food; and removing, washing, and drying the plates.*

(2) *The lesson included (1) cooking the steak, eggs, and chips, (2) adding salt, pepper, and sauce, (3) tasting the food, and (4) removing, washing, and drying the plates.*

4.5 To separate "etc." from list items:

in general, a *comma* or *semicolon* is needed

Punctuate as if "etc." is replaced by "and so forth"; thus a comma is needed if your preference is to use a comma before *and* in a list (i.e., a serial comma—see section 4.2 above).

Example:

The restaurant stocked only New World wines (from Australia, California, Chile, South Africa, etc.), but the selection was excellent.

5 Show a possessive relationship

5.1 To indicate a possessive noun:

use an *apostrophe*.

The general rule is to add *'s* to a singular noun (e.g. 1–3 below) and just an apostrophe to a plural noun (e.g. 4 and 5). However, when a plural noun does not end in an *s*, add *'s* (e.g. 6 and 7). Exceptions arise with personal names—see **Grammar and style [Possessives]**.

Examples:

(1) *the Earth's climate*

(2) *the water's salinity*

(3) *the writer's assumption* (i.e., one writer)

(4) *the holes' diameters* (i.e., many holes)

(5) *the boxes' lids* (i.e., more than one box)

(6) *the chairmen's notes*

(7) *the women's race*

5.2 To indicate a possessive pronoun:

in general, *no apostrophe* is used.

To indicate a possessive construction most pronouns change their form and are written without an apostrophe—for example: *she/her* (e.g. 1 and 2 below), *they/their* (e.g. 3 and 4), *we/our* (e.g. 5 and 6), and *you/your* (e.g. 7 and 8). But a few pronouns do not change form and need an apostrophe (e.g. 9 and 10).

Examples:

(1) *her result* (2) *it was hers*

(3) *their test* (4) *it was theirs*

(5) *our expectations* (6) *they were ours*

(7) *your answers* (8) *they were yours*

(9) *one's methodology* (10) *anyone's guess*

6 Quote something (see also **Quotations**)

6.1 To indicate a quotation:

use *quotation marks.*

Written or spoken words attributed to someone else are enclosed in quotation marks (e.g. 1 below). When the quotation is introduced by what is a complete sentence a colon is needed (e.g. 2). A colon may also be used to start off a long quotation (i.e., a long sentence or multiple sentences) (e.g. 3).

The closing quotation mark is always placed after a comma or period (e.g. 1–3), but before a colon or semicolon (British English convention is different—see **Quotations**). A question mark (or an exclamation point) is placed inside the quotation marks if it relates to the quoted words (e.g. 4) and outside if it relates to the enclosing sentence (e.g. 5).

Examples:

(1) *"Unsolicited proposals," according to the briefing document, will not be considered.*

(2) *The request was explicit: "The duration of the project will be three years."*

(3) *The e-mail from the Commission stated: "Your proposal scored 7.3 out of 10. It was ranked sixth and has been placed on the reserve list for funding."*

(4) *The question "What are the novel aspects of the proposal?" must be answered in a way that does not violate Intellectual Property agreements.*

(5) *What is meant by "project life-cycle cost"?*

6.2 To leave something out of a quotation:

use *ellipsis marks.*

The part of a quotation that is omitted is replaced by ellipsis marks (see example below), which some authors like to put in brackets.

Example:

In the small print, it was indicated that "conflicts of interest ... must be declared."

6.3 To insert a comment into a quotation:

use *brackets.*

Brackets are used when a comment is inserted into a quotation (e.g. 1 below). By convention, brackets are also used to enclose the Latin word *sic*, which is used to indicate an error in quoted material, such as a spelling mistake (e.g. 2).

Examples:

(1) *The proposal "must be submitted electronically [to the Commission] using the supplied template."*

(2) *It was requested that "the principle [sic] investigator" be indicated on the front cover.*

6.4 To identify quoted or emphasized words within a quotation:

use *single quotation marks.*

Use single quotation marks, rather than the standard double quotation marks, to identify quoted or emphasized words within a quotation.

Example:

The evaluators were of the opinion that the project "failed to address the 'environmental concerns' that were mentioned in the Request For Proposal."

7 Clarify word usage

7.1 To indicate that two or more words are to be understood as forming a single adjective:

use *hyphens.*

A compound phrase used as an adjective before a noun should be hyphenated (e.g. 1 below), but an adverb modifying an adjective does not require a hyphen (e.g. 2). (See also **Hyphenation and word division** for further details on the use of the hyphen.)

Examples:

(1) *It was a typical back-of-the-envelope calculation.*

Example of incorrect usage:

(2) *The overly-optimistic calculation predicted significant performance improvements.*

7.2 To join a prefix to a proper name or add a prefix or suffix to a symbol or numeral:

use a *hyphen.*

Examples:

(1) *his position was anti-Einsteinian*

(2) *an I-shaped cross section*

(3) *a 10-fold increase*

7.3 To indicate a jargon word or phrase:

if required, use *quotation marks.*

Technical jargon is often unavoidable in reports. You can identify jargon words and phrases by quotation marks if you wish (e.g. 1 and 2 below), but in most cases it is not necessary.

Examples:

(1) *The computer technician "daisy-chained" the drives.*

(2) *A "fudge factor" of 1.2 was used as a multiplier to account for the increased component size.*

8 Indicate a range or link

8.1 To indicate a range of numbers or specify a time period:

use an *en dash* (an en dash is longer than a hyphen).

Use a dash to indicate a range of numbers (e.g. 1 below) including a time period (e.g. 2). It is not necessary to repeat the units (e.g. 3), but the % sign is usually repeated in a range (e.g. 4). As the dash is read as *to* do not write *between* before the range (e.g. 5). (See also **Numbers [Ranges]**.)

Examples:

(1) *pp. 179–206*

(2) *1995–98*

(3) *80–120 μm*

(4) *50%–75%*

Example of incorrect usage:

(5) *between 1995–98*

8.2 To link two places or two people:
use an *en dash.*

Examples:

(1) *Seattle–Vancouver railway*

(2) *Mann–Whitney test*

8.3 To indicate a link between two associated terms (typically two nouns) that have a relationship or connection:
use an *en dash* or a *slash.*

Examples:

(1) *stress–strain curve*

(2) *elastic–plastic transition*

(3) *volume–volume ratio* or *volume/volume ratio*

(4) *P/E multiple*

(5) *go/no-go gage*

Punctuation marks and special symbols/characters

A list of commonly encountered punctuation marks and special symbols is given in the table below.

Punctuation marks and special symbols/characters

<table>
<tr><th>Mark</th><th>Name</th><th>Notes</th></tr>
<tr><td>.</td><td>period, full stop, full point, dot</td><td>full stop is British English; dot is used in URLs.</td></tr>
<tr><td>?</td><td>question mark</td><td></td></tr>
<tr><td>!</td><td>exclamation point, exclamation mark</td><td>exclamation mark is British English</td></tr>
<tr><td>,</td><td>comma</td><td></td></tr>
<tr><td>:</td><td>colon</td><td></td></tr>
<tr><td>;</td><td>semicolon</td><td></td></tr>
<tr><td>...</td><td>ellipsis marks, ellipsis points</td><td></td></tr>
<tr><td>’</td><td>apostrophe</td><td></td></tr>
<tr><td>-</td><td>hyphen</td><td></td></tr>
<tr><td>–</td><td>en dash, en rule</td><td>dash of approximate width of the letter N</td></tr>
<tr><td>—</td><td>em dash, em rule</td><td>dash of approximate width of the letter M</td></tr>
<tr><td>“ ”</td><td>double quotation marks, inverted commas</td><td>often just called quotation marks</td></tr>
<tr><td>‘ ’</td><td>single quotation marks, inverted commas</td><td></td></tr>
<tr><td>( )</td><td>parentheses, round brackets, brackets</td><td>brackets is British English</td></tr>
<tr><td>[ ]</td><td>brackets, square brackets</td><td>square brackets is British English</td></tr>
<tr><td>{ }</td><td>braces, brace brackets, curly brackets</td><td></td></tr>
<tr><td>< ></td><td>angle brackets</td><td></td></tr>
<tr><td>« »</td><td>chevrons, guillemets</td><td>guillemets are quotation marks in French</td></tr>
<tr><td>/</td><td>slash, forward slash, solidus, virgule</td><td></td></tr>
<tr><td>\</td><td>back slash</td><td></td></tr>
<tr><td>|</td><td>vertical line, vertical bar</td><td></td></tr>
</table>

Punctuation marks and special symbols/characters (*continued*)

Mark	Name	Notes
_	underscore	
′	prime mark	
"	ditto marks	
*	asterisk	
&	ampersand	
@	at sign	
%	percent *or* per cent	per cent is British English
‰	per mill	means *per thousand*; its use is not recommended
#	hash, pound, or numeral sign; octothorp	
~	tilde	used for approximate values
©	copyright sign	
®	registered sign (i.e., trademark registered)	usually written as superscript
™	trademark sign	usually written as superscript
§	section mark	
¶	paragraph mark, pilcrow	

Notes: The less commonly used symbols and marks, which do not appear on a standard keyboard, are obtained in Microsoft Word by using the Insert pull-down menu and selecting Symbol [Special characters]. For many of these characters there are shortcut keys—for example: Alt-0150 on the numeric keypad gives an en dash and Alt-0151 an em dash.

Q

Quotations

A quotation is an exact copy of words from another work or a transcription from a speech, interview or dialogue. Make it clear to readers that these are not your own words, and reference the quotation appropriately. A long quotation (typically more than 40 words) should be treated as an *excerpt*, or block quotation.

Quotations: rules and conventions

- Quoted words, sentences and short passages should be enclosed in quotation marks when included in the line of the text. Example (1): *King (2005) reported that "low-carb French fries can be made by replacing most of the potato with cauliflower, egg, and calcium caseinate"; however, no mention is made of the taste.*
- Do not change the spelling (say from British to American English), punctuation, or word sequence (it would be incorrect to change the informal *low-carb* to *low-carbohydrate* in the above example). It is, however, acceptable to change the initial letter of the first word from a capital to a lowercase, or vice-versa, to blend into the passage. Similarly, the closing punctuation may be changed provided that the meaning is not altered (a question may be turned into a statement by omitting the question mark, for example).
- Mistakes in the quotation (i.e., things that the person being quoted got wrong) may be identified by inserting the word *sic* in brackets after the mistake. If, however, the mistake is somewhat obscure and there is a high probability that readers will not know what the correct version should be (e.g., an incorrect date or spelling of a scientific term) then it would be better to give the correct version in brackets. Example (2): *"Beta-carotene, one of the best-known photochemicals [phytochemicals], gives color to carrots and other orange, yellow, and red produce."*

- Text omitted from a quotation should be replaced by ellipsis marks (i.e., three periods). Example (3): *Le Boeuf*[(17)] *is of the opinion that a "high-protein, high-fat diet ... represents a significant health risk to children."*

- Comments or interpolated words inserted into quotations should be enclosed in brackets. Example (4): *Le Boeuf*[(17)] *continues: "The mood-enhancing effect of fat was shown [in clinical trials] to be similar to that of sedative-hypnotic drugs."*

- A direct quotation that is a complete sentence or two can be introduced by a colon (as described in **Punctuation** section 6.1), but that is not the only acceptable way to do it: some authors prefer to use a comma (e.g. 5), while others just open the inverted commas. Example (5): *In his opening address, Edward King said, "Potatoes contain glycoalkaloids, which, in rare cases, causes headaches and diarrhea."*

- Punctuation rules for American and British English differ when the closing quotation mark coincides with a punctuation mark related to the enclosing sentence:

  (a) In American English the closing quotation mark is always placed after a comma or period (see e.g. 6), but before a colon or semicolon. Example (6): *The research brings new meaning to the phrase "addicted to fast food."* A question mark or exclamation point is placed inside the quotation marks if it relates to the quoted words and outside if it relates to the enclosing sentence. Example (7): *What exactly are "sedative-hypnotic drugs"?*

  (b) British English convention requires that only punctuation related to the quoted words be included as part of the quotation and that punctuation related to the containing sentence be placed outside of the quotation marks. Thus, only when the quotation is a complete sentence requiring a period is it placed inside the quotation marks, and there is no need for an additional period after the closing quotation mark (see e.g. 5 above, which conforms to both American and British English conventions).

- If the source (from which you are quoting) contains a reference to another work that is not regarded as significant (for your report), it may be omitted. The details of cited references in quotations that are retained should be given after the quotation using a different typeface or smaller font size.

- Long quotations from publications protected by copyright, even if correctly referenced, can get you into trouble. The purpose of the copyright is to give an exclusive right for certain uses of the work to the holder (see **Copyright**); however, under the so-called **fair-use doctrine**, a limited amount of copyright material may be reproduced without permission.

Excerpts (block quotations)

- Excerpts are long quotations; they should be formatted as a separate paragraph (or paragraphs) indented from the left margin of the encompassing text. Quotation marks are not required, but a change of font size (say from 12 to 10) is sometimes used to distinguish the excerpt from the surrounding text.

- If the source is not identified in the text preceding the excerpt, it should be stated immediately after the closing punctuation of the excerpt.

See **Citing references (basic rules)** and **Plagiarism**.

R

Reader

It is obvious that a report is written for someone else to read; however, if this point is not taken seriously, the report can end up as a personal diary: clearly understandable to the writer, but baffling to the reader. It does not matter if the reader is your long-suffering professor, your boss, or a client; you need to consider how that person will view what you have written.

A common mistake is to structure a report in a way that mirrors the chronological sequence of the work. This is usually not the best approach, although it may initially appear to be the simplest. Place yourself in the reader's position—remember that he or she may have little or no knowledge about your particular study—and then work out a logical structure that will best convey the whole picture.

Help the reader navigate by providing clear signposts (e.g., page and chapter numbers, appropriate chapter and section headings) and by good **cross-referencing** (to figures, tables, and other parts of the report, including the appendix). Note that a reader with some knowledge of the subject may start with the **abstract** and **objectives**, scan a few chapters to get an overall impression of the work, and then read the **discussion** and **conclusions**, before coming back to read selected chapters (provided that he or she has not lost all interest by then and used the report to stoke a fire).

A question that often gets asked is: *What can I assume the reader already knows and I therefore don't have to explain?* In the absence of any concrete information about the target audience, write the report for a peer who has a technical knowledge comparable to that which you had before you conducted the study that you are now writing about.

Recommendations

- The recommendations may be written together with the **conclusions** under one chapter heading or written as a separate chapter, located after the conclusions (which is usually a superior approach). Popular alternative headings for this chapter are *future work* and *future research* (see **Future work**).

- The recommendations are best presented as a list—either bulleted or numbered. The latter option has the advantage that a reader can refer to a specific recommendation.

- All recommendations should be unambiguously stated and drawn directly from the work conducted.

- It is rare that a single study will answer all possible questions related to the subject reported on. The constraints and limitations imposed on the study—by the tools used (hardware and software), time schedule, input data, availability of specimens, and so forth—invariably mean that further work is possible. Provide focused recommendations in your report for future work, identifying issues that, in your opinion, would lead to the biggest payback if follow-on investigations were to take place; suggest ways for getting more accurate or complete answers and indicate opportunities for the exploitation of your results.

- In certain engineering studies, the recommendations would be the most important aspect of the work—for example, following the assessment of the physical condition of a construction (such as the roof of a seventeenth-century cathedral) the recommendations would be for a specific course of action. Similarly, an engineering team, commissioned to evaluate a new prosthesis design, for example, may recommend in their report safety-related improvements to the design concept.

References (basic rules)

- The reference list starts on its own page and is located after the last chapter of the report, but before the appendix.
- Each reference that you have cited in the text should have an entry in the list. Do not include references that are not cited in the report. If you wish to produce a list of relevant publications in the field of study, compile a **bibliography**.
- There are a number of citation styles (e.g., APA, CSE, Chicago Manual, MLA, Turabian), each with its own set of rules governing the format and punctuation of publication details (academics tend to make a big deal about the differences, but in substance they are all much the same). The important issue is that you provide sufficient information so that readers can locate the exact work you consulted. The precise format and punctuation are a matter of personal choice.
- The citation styles fall into three categories—see **Citing references (basic rules)**—two of which require the generation of a reference list: the *author-date method* and the *numeric method.* The third category, which utilises footnotes or endnotes, would typically involve the generation of a bibliography (not popular in engineering).
- The required information and the sequence appropriate to the *author-date method* are given in the table below. For the *numeric method* there is one significant difference: the year of publication does not follow the author—the year or publication date (if available) is placed after the publisher (or equivalent).

Required reference information

Reference work	Information required
book	Author(s). Year[1]. *Title of book.* [Edition/volume.] Place of publication: publisher.
titled chapter/ section in an edited book	Author(s) of chapter. Year[1]. Title of chapter. *Title of book.* [Edition/volume.] Chapter number/page numbers. Editor(s). Place of publication: publisher.

Required reference information (*continued*)

Reference work	Information required
journal paper (article)	Author(s) of paper. Year[1]. Title of paper. *Title of journal.* Issue information (i.e., volume number, issue number, month, season). Page number(s) of paper.
proceedings (e.g., conference paper)	Author(s) of paper. Year[1]. Title of paper. *Conference title.* Location and date of conference. Page number(s) of paper. [Editor(s). Place of publication: publisher.]
report	Author(s). Year[1]. Title of report. [Version/edition.] [Reference number.] Name of organization. Abridged address. Date of publication. [Distribution restrictions.]
thesis or dissertation	Author. Year[1]. Title. Degree award. [Name of supervisor.] Name of university. Abridged address.
patent	Author(s)/originator(s). Year[1]. Title. Designation/ reference number. Date.
newspaper, magazine, or newsletter article	Author(s). Year[1]. Title of article. *Title (name) of newspaper/magazine/newsletter.* Date. Page number(s) of article.
website	Author(s). Year[1]. Title of document. [Version/ reference number.] [Volume/section/page number(s).] [Name of organization.] URL [Date of "publication."] Date of access.

Notes:

1 The indicated sequence is for the author-date method. For the numeric method the year of publication does not follow the author(s); the year (or date) is placed after the publisher (or equivalent).

2 Details in brackets may not be required or applicable.

3 A work that has more than one author should have the co-authors listed in the order given in the original work—the family name of the lead author comes first, followed by his/her initials (or forenames) and then the other authors. If there is a long list of authors, say eight or more, then only the first six names are listed, and the seventh and subsequent authors are abbreviated as *et al.* Degrees, titles, and affiliations (of professional bodies, for example) following the names are omitted.

4 The titles of published works are traditionally identified using italics or by underlining: as hypertext links are underlined, italicization is recom-

mended. But be careful here: this refers to the title of the book or journal and not the title of the chapter or article.

5. Use the full title or an accepted abbreviation for a journal (see **Abbreviations (of journals)**).
6. Abbreviations (such as *vol.*, *chap.*, and *p.*) should be used to define a part of a reference work; however, if it is clear what the numbers designate, then the abbreviations may be omitted.
7. For the place of publication just indicate the city. If this is not a well-known place or if there is more than one city with that name, add the state or country.
8. If you cannot find a piece of information, or it is not applicable, leave that part out; but give as much relevant information as you can find.
9. Following the publication details, a statement qualifying the availability of the document may be added—for example, terms such as *Unpublished, Confidential report, Restricted distribution* are used. Similarly, reference works that are not in print format usually have a medium descriptor afterwards (in brackets or parentheses)—for example: *[videocassette]* or *[CD]*.

References: publication date

- ❑ The author-date method requires that the second element of information be the year of publication. There are, however, two exceptions: (1) a work that has no date should have the abbreviation *n.d.* in place of the year and (2) a work that has been accepted for publication, but not yet printed, should have the words *in press* in place of the year. This information should always match the "date" in the citation in the text, as it links the citation with the corresponding entry in the reference list.

- ❑ For the numeric method (and for bibliographies) the date usually follows the publisher (or equivalent); it can be the year of publication (e.g., in the case of a book) or the month and year (e.g., for a monthly periodical) or an exact date (e.g., for a published report).

References: punctuation, format, and typography

- ❑ Unless you are following a particular citation style (e.g., APA, CSE, Chicago Manual, MLA, Turabian), it makes no difference whether you record the author's forename(s) or just his or her initials; whether you put the year in parentheses or not; or whether the individual elements of information are separated by commas, semicolons, or periods. These are issues of style. It is, however, important that you provide complete details for every reference in a consistent manner.

- ❑ If you have a long list of alphabetically ordered references, select a format that will make it easy for readers to find a particular entry—for example, by setting the name of the lead author in bold font or, alternatively, by indenting the second and subsequent lines of each entry (i.e., a hanging indent) to make the lead author more prominent.

Reference list entries: examples

See **References (examples of author-date method)**, **References (examples of numeric method)**, and **References (non-archival)** for examples. A detailed treatment of the subject, with examples conforming to individual citation styles, is provided in the works listed under **Style manuals**. Additional information, specific to the Internet, is given in **Internet reference citation**.

References (examples of author-date method)

Sequence of entries in the reference list

The references are listed in alphabetical order by the family name of the lead author and are not numbered or divided into sections. The following rules apply to sort out the sequence of entries:

- Works that have exactly the same author (or authors) are listed chronologically (the earliest work is first).
- Works that have exactly the same author (or authors) and were published in the same year are distinguished by the addition of a lowercase letter (a, b, c, etc.) after the year; these letters are allocated in the order in which the works appear in the text. This applies to the citation in the text—for example: *Phoolproof*

(2001a) developed a program to analyze the loads—and to the corresponding entry in the reference list.

- When the same lead author appears with different co-authors, the entries are listed alphabetically according to the names of the co-authors.
- A single-author entry is listed before a multiple-author work with the same lead author.

References: book

Muffet, L. M., Curd, T. P., and Whey, F. 2004. *Treating Arachnophobia: Theory and Practice*. Sydney: Terror Press.

Phoolproof, E. 2001a. *Spacecraft Dynamic Loads*. Vol. 1. Roswell, NM: Cosmic Ocean.

Phoolproof, E. 2001b. *Spacecraft Dynamic Loads*. Vol. 2. Roswell, NM: Cosmic Ocean.

Notes:

1. You can give the authors' forenames as they appear on the title page (i.e., not replaced by initials), if you wish.
2. A stylistic quirk of the *Chicago Manual of Style* is that the family–forename order is reversed for the second and subsequent authors. If the forenames in the above example were written out, the entry would appear as: Muffet, Loretta M., Tiffany P. Curd, and Francis Whey. (This can result in a jumble of initials after the lead author.)
3. The little *a* and *b* after 2001 for the two references of Phoolproof illustrate the accepted solution to the problem of having two (or more) references by the same author with the same year. (The *a* and *b* would also follow the year in the text citation.)

References: edited book (referring to the whole book, not just a section)

Cole, H. M., ed. 1998. *Favourite Violin Classics for Three Players*. 2nd ed. London: Mother Goose Press.

Solomon, A. T. P. and Grundy, M. M., eds. 2005. *Modern Science of Longevity*. Los Angeles: Snake Oil Pub.

References: titled chapter/section in an edited book

Whitethorn, K. F. 2000. Chopping Down the Fairy Tree. In *Modern Collection of Irish Short Stories*, pp. 137–145. Murphy, P. J., ed. Cork: Blarney Books.

References: journal paper (article)

Dish, D. A. and Spoone, F. G. 2003. Investigation Into Bovine Anti-gravity Techniques: Part 1 Theoretical Formulation. *Ph. Sci. Rev.* Vol. 13, 2003, pp. 167–171.

Dish, D. A. and Spoone, F. G. 2004. Investigation Into Bovine Anti-gravity Techniques: Part 2 Experimental. *Ph. Sci. Rev.* 16(2004), 208–213.

Sparrow, L. B. J. In press. Failure Modes of Birdlime. *International Journal of Glues and Adhesives.*

Notes:

1 You can use either the full title or an accepted abbreviation for the journal (see **Abbreviations (of journals)**).

2 The numbers *16(2004), 208–213* (used in the second example) stand for *volume 16 of 2004, pages 208 to 213.* Both formats (illustrated in the two examples of Dish and Spoone) are acceptable, but it is important to be consistent in the selected approach.

3 The words *in press* (in the third example) indicate that the work has been accepted for publication, but has not yet been printed.

References: electronic journal (e-journal) paper (article)

Mudd, J. J. 2002. Review of Traction Enhancement Techniques. *E-Journal of Mechanized Farming.* Vol. 8, 2002, pp. 11–12. http://www.prof_e-journals.edu/mechfarm_mar02/mudd.html. Accessed May 18, 2003.

Notes:

1 A distinction is made between archival publications that are available on the Internet (such as journals, books, conference proceedings, and so forth) and run-of-the-mill personal or company websites. For the latter

category, it is possible that readers would not be able to access the exact work that you did (as web pages are changed quite often); for this reason these websites are classed as *non-archival* reference sources—see **References (non-archival)** and **Internet reference citation**.

2 See **URL** for details on writing the Internet address.

References: proceedings (of a conference, meeting, or symposium)

Noonan, L., O'Callaghan, T. I. P., and Milligan, R. L. 2004. Jet Engine Noise and its Impact on Passing Cyclists. *Second European Symposium on the Environmental Impact of Modern Transportation Systems*, Dublin, April 1.

Hubbard, M. S. 2002. Calcium Deficiency in *Canis familiaris*—A Modern Problem. *Tenth International Domesticated Animal Health Conference*, New Orleans, Dec. 12–13, pp. 124–127. Proceedings edited by Petz, E. R., Atlanta: Southern Academic Press.

Note: The date of the symposium/conference has been shortened and does not include the year, which is given after the authors' names. (The same idea applies to reports, newspaper articles, personal correspondence, and so forth when the date is specified.) This approach is recommended in most **style manuals** (the objective is to make the reference as concise as possible); however, there is no compelling reason why the year cannot be repeated as part of the date. As is the case with all such conventions, it is important to be consistent throughout the report.

References: report

Hansell, J.-P. and Gretal, R. T. 2003. GPS Performance Reduction due to Dense Forest Canopy. Technical report KL-41. GlobalView Consulting, 321 Hexereistrasse, Freiburg, Germany, Oct. 31.

References: thesis/dissertation

Ples, M. S. 2005. An Investigation into Parallels of Palaeolithic Nutrition and Modern Fast Food Diets. Ph.D. diss., University of West Sterkfontein, South Africa.

References: patent

Piper, P. 1986. Aural Rodent Lure. European patent application 0021168 A2. 1986-01-09.

References: magazine, newspaper, or newsletter article

Princess, P. 2004. Insomnia: Too Much Caffeine? *New England Chronicle*, July 30, p. 15.

Green Bay Examiner. 2005. Editorial. Feb. 14, city edition.

Note: When the author is not indicated (as is the case in the second example) the name of the organization serves as the "author" in the text citation and in the reference list.

References: database on the Internet

NIST. 2003. NIST Chemical WebBook: Standard Reference Database No. 69. National Institute of Standards and Technology (NIST), Gaithersburg, MD, USA. http://webbook.nist.gov/chemistry/. Release: March 2003. Accessed Nov. 18, 2004.

References (examples of numeric method)

Sequence of entries in the reference list

There are two ways of numbering the references (and this dictates the sequence):

(1) The references are allocated consecutive numbers according to the order in which they first appear in the text. In the reference list they are recorded in the order of their citation (called *citation-sequence* style).

(2) The references are ordered alphabetically by the family name of the lead author and then allocated consecutive numbers. In this case the numbers will not be sequential in the text (this option is seldom used).

References: position of the date

The order in which the reference details are provided is the same as that of the author-date method—see **References (basic rules)**—with the following exception: the year is not placed after the name(s) of the author(s); instead, the year (or complete date) appears after the details of the publisher (or equivalent).

References: book

[5] Skywalker, M. A. D. *Aircraft Design: Getting it Right so you Don't Crash on Takeoff.* 2nd ed. Dayton: Kittyhawk Pub., 2003.

References: edited book (referring to the whole book, not just a section)

13 Foster, D. R., ed. *Waterproof, breathable expanded-PTFE fabrics.* Gloucester, UK: Soggy Books, 2005.

References: titled chapter/section in an edited book

9\. Ironside, H. G. Corrosion Prevention and Control. In *Synthesis of Modern Ship Hull Design*, pp. 364–389. Loftsman, J. K., ed. Liverpool: Barnacle and Son, 2000.

References: journal paper (article)

(2) Pike, O. G. and Chipps, K. B., Jr. Comparative Attention Deficit Disorder in *Carassius auratus*. *Journal of Ichyology and Ichthyosis*. Vol. 15, 2002, pp. 102–103.

(16) White, S. Maximizing Yield in Low-grade Diamond Ore. *J. Mat. Proc.* 24(2003), 119–122.

Notes:

1 Use an accepted abbreviation for the journal title if you wish (see **Abbreviations (journals)**).

2 The numbers *24(2003), 119–122* imply *volume 24, year 2003, pages 119 to 122.*

References: proceedings (of a conference, meeting, or symposium)

14 Undulait, B. N. Multivariate Optimization of Water Turbine Design. *Fifth International Wave Energy Conference*, Alice Springs, Australia, Jan. 13–15, 2005, pp. 189–197.

22 Bukkel, P. and Burste, C. M. Optical Fiber Strain Measurement: A Review. *European Smart Composites Forum*, Berlin, Nov. 19–21, 2002. Proceedings in: *Notes on Numerical Analysis*, Vol. 76, 2003, pp. 269–277, Zurich: Starlight-Little.

References: report

3. Hood, L. R. R. *The Treatment of Eating Disorders in Canis lupus*. Report HY-173. Department of Applied Animal Psychotherapy, Grannie College, Woodcut Valley, WA, USA, n.d.

Note: The abbreviation *n.d.* means *no date*.

References: thesis/dissertation

[52] Post, A. N. Investigation Into the Influence of Coronal Mass Ejections (CMEs) on the Homing Abilities of Carrier Pigeons. Ph.D. thesis, Columba Livia Institute of Technology, New Brunswick, Canada, 2002.

References: patent

(17) Sapien, H. Detachable Tags for Y-chromosome Mounts. US Patent No. 5-2088, 456, Feb. 28, 1969.

References: magazine, newspaper, or newsletter article

7. Pattern, L. S. Global warming: opportunity or calamity? *Science&Money*. Oct. 12, 2004, pp. 31–35.

8. *Melbourne Inquirer*. Editorial. June 15, 2005, evening edition.

Note: When the author is not indicated (second example) the name of the organization serves as the "author."

References (non-archival)

Non-archival references are published and unpublished reference sources that are not routinely stored (by libraries and government agencies, for example) and made available for widespread distribution. This would include brochures, pamphlets, product data sheets, newsletters, and so on. Most websites are classed as non-archival reference sources as there is no guarantee that the information will not change soon after the site is accessed (see **Internet reference citation**). On the other hand, archival references are books, journals, conference proceedings, newspapers, and many scientific/technical reports (i.e., the material that libraries routinely catalog and make available on request).

Avoid non-archival references if there is another source of the same information. If you use a non-archival reference, include as much relevant information as you can about the source, such as author(s), date, title and name, and address of organization. If there is no author, it is acceptable to write *Anon.* (for anonymous); however, if you are using the author-date method, it is better to use the organization name (or an abbreviation or acronym, if one exists) as the "author."

Restrictions regarding distribution or confidentiality clauses associated with the document should be indicated after the publication details.

References: company or organization document

Author-date method:

Penn, S. T. E. 2003. Internal Report on the Pilfering of Stainless Steel Roller Balls. Last Legs Engineering Inc., Willie Sutton Rd., Toronto, Ontario, M5S 3G8, Canada, Apr. 19. [Confidential report].

Numeric method:

[6] Blankett, W. E. T. (Chairperson). The Safety Challenge. Report of the Technology Sub-Group, Study Conducted Under the Aegis of the Department of Enterprise, Dec. 2002. Available from National Society of Go-cart Builders, Hamilton Place, London W1V OBA, UK.

References: anonymous document with limited distribution

Author-date method:

SGWTC. 2003. French Add Acid! Autumn Newsletter of the Sour Grapes Wine Tasting Club (SGWTC), 14 New Alsace St., Lake Shallow, NSW, Australia, Apr.–Jun.

Numeric method:

(19) Anon. French Add More Acid!! Winter Newsletter of the Sour Grapes Wine Tasting Club (SGWTC), 14 New Alsace St., Lake Shallow, NSW, Australia, Jul.–Sep. 2003.

References: website

Author-date method:

ICAO. 2004. Airport Charges. International Civil Aviation Organization (ICAO), Montreal, http://www.icao.int/. Accessed May 12, 2004.

Numeric method:

(11) ICAO. Airport Charges. International Civil Aviation Organization (ICAO), Montreal, http://www.icao.int/. Accessed May 12, 2004.

Notes:

1 Referencing documents accessed via the Internet can be problematic—see **Internet reference citation**.

2 See also **URL** for details on writing the Internet address.

References: personal communication

Author-date method:

Moreau, J. 2005. Personal correspondence with Dr. Jacqueline Moreau, Senior Research Officer, Widgets Prosthesis, 165 rue de Deloire, 75008 Paris, Jan. 8–10.

Numeric method:

[4] Moreau, J. Personal correspondence with Dr. Jacqueline Moreau, Senior Research Officer, Widgets Prosthesis, 165 rue de Deloire, 75008 Paris, Jan. 8–10, 2005.

References: e-mail

E-mail is personal correspondence and can be referenced in the format shown above; alternatively, and this is recommended in a number of **style manuals**, the sender's e-mail address is given:

Author-date method:

Segantini, A. 2004. Personal e-mail sent from Mr. Antonio Segantini to the author. Subject: Your request concerning materials data. a.g.segantini2002@freemail.com. Sent Sep. 12.

Numeric method:

(8) Segantini, A. Personal e-mail sent from Mr. Antonio Segantini to the author. Subject: Your request concerning materials data. a.g.segantini2002@freemail.com. Sent Sep. 12, 2004.

Repetition

It is correct—and necessary—for you to repeat what you say in the report, as this is called for by the traditional report structure. For example, mention of the experimental technique that was used could be made in the **abstract** (keywords only), **introduction** (to set the scene), **method** (to record precisely what was done), **discussion** (to interpret or to discuss the process and the results), and **conclusions** (to assess the process and results). Similarly, important results could be described in the **abstract**, **results**, **discussion**, and **conclusions**. Note that the purpose is never the same; so, although you are writing about the same issue, the style and emphasis are not identical in each chapter.

Reports are seldom read as novels from cover to cover—readers may scrutinize the **abstract** and **objectives**, skim through a few chapters to get a flavor of the work (possibly ignoring the **literature review** and **background** sections), study the **discussion** and **conclusions** (not necessarily in that order), and then, if the work is of interest, they may start again at the beginning. The traditional report structure is a proven framework that enables readers (with different interests and backgrounds) to get a sensible overview of the work during the first pass through the report and then, on the second pass, to explore it in greater detail. The framework necessitates a structured repetition of the critical elements, but this requirement should never be used as an excuse to bump up the thickness of the report by mindless cutting and pasting of paragraphs from one chapter to another.

Results

- The results chapter is an unadulterated record of the relevant experimental and/or calculated data in **tables** and **figures**, explained by simple factual statements. Never manipulate results to suit your image of what they should look like (see **Bad results**); rather present them accurately—and in the **discussion**, be honest in your interpretation of the results; identify gaps and apparent inconsistencies in the data.

- Be careful when you include the results of other researchers in your report (say for comparative purposes)—you need to clearly indicate the origin of these results so that readers are never left wondering what was done by whom.

- Do not dump the results into the report without any explanation: draw the reader's attention to the data presented (remember that every table and figure should be mentioned in the text). It is acceptable, and often very helpful to the reader, to highlight trends or observations; but detailed analyses and discussions should not be presented here, as they belong in a later chapter: the **discussion**.

- Do not include long lists (or tables) of raw data in the results chapter (unless you deliberately want to bore the readers to tears); they should be placed in the **appendix**. Rather include a summary of the data and an explanation of what was done. Indicate the precise location of the complete data set (i.e., the table/figure number or page number in the appendix) and explain how the data was manipulated or summarized.

- Detailed analyses—not regarded as critical for an understanding of the results presented—should also be placed in the appendix. However, you must include all *important* results of the study in the results chapter (which can be repeated, if necessary, in the appendix). Readers should never have to hunt for a vital answer that you have buried in a table in the appendix (place yourself in the reader's position and imagine how frustrating that is for someone who is tired and has 10 more reports to read!).
- Typical structure of the results chapter:
  - Review the methodologies used for collecting and manipulating the data.
  - Present the results in **tables** and **figures**. It is not necessary to present the data twice (i.e., in a table and a figure). If the precise numbers are less important that the trends, just use a figure. Associated with each table or figure should be a short description, highlighting the important elements and permitting the readers to quickly interpret the data.
  - Describe the data analysis undertaken (if appropriate)—for example, statistical analysis may be used to determine whether or not an observed difference or correlation is significant.
- See also **Data presentation**, **Statistical analysis**, and **Error analysis (measurements)**.

Revision control

- Revision control (sometimes called configuration control) is a document management process—used in industry, government departments, scientific research agencies, and so forth—to ensure that, when reports are updated or superseded, all users have the most recent version. Its main purpose is to provide a traceable record of updates made to the report.

- Adopt the following guidelines for reports requiring revision control:
  - Put a date on the report, corresponding to its release/approval.
  - Give the report a reference number and a revision (version) number or letter.
  - Number all pages and identify the revision number or letter on each page.
  - Create a "revision sheet," which would list the updated or superseded pages, the dates of the changes, and, optionally, the signatures of the persons responsible for approving/implementing the changes. This is frequently printed on colored paper and placed after the title page.
  - Keep an up-to-date list of recipients of the document.

Roman numerals

Roman numerals

Roman	Arabic	Roman	Arabic	Roman	Arabic
I	1	XI	11	L	50
II	2	XII	12	C	100
III	3	XIII	13	D	500
IV	4	XIV	14	M	1 000
V	5	XV	15	$\overline{V}$	5 000
VI	6	XVI	16	$\overline{X}$	10 000
VII	7	XVII	17	$\overline{L}$	50 000
VIII	8	XVIII	18	$\overline{C}$	100 000
IX	9	XIX	19	$\overline{D}$	500 000
X	10	XX	20	$\overline{M}$	1 000 000

S

Scientific names (of animals and plants)

- Rules of taxonomy are complex—for details consult *Scientific Style and Format: The CSE Manual for Authors, Editors, and Publishers* (Rockefeller University Press, 2006), for example.
- When writing scientific (Latin) names of animals and plants, note that
  - The names of the genus, species and subspecies (if applicable) are italicized.
  - The initial letter of the genus name is capitalized, but the species and subspecies are written in lowercase (e.g., *Australopithecus africanus*).
  - Divisions higher than genus (e.g., phylum, class, order, and family) are not italicized.
  - The genus name may be abbreviated after its first mention (e.g., *A. africanus*). However, some things are better known with the genus abbreviated (e.g., *E. coli*); in such cases there is no value in writing out the genus name when it is first mentioned.

Sections in a report

See **Chapters and sections** and **Structure of reports and theses**.

Significant figures

- In most cases, the number of significant figures is determined by counting the digits of a number starting from the first non-zero digit and ignoring trailing zeroes—for example: the number 0.027 010 (which can be written in scientific notation as 2.701×10^{-2}) has four significant figures. Similarly, the number 3 490 200 has five significant figures (although it is only possible to be sure if the number is written in scientific notation).

- There is an implicit assumption that the number of significant figures of a result given in a report relates to the accuracy of the measurement or calculation undertaken. In a design study of a new racing boat, for example, the answer: *The engine power was estimated to be 1385.968 kW*, would show poor understanding of this concept. To give the calculated power of a large turbine engine to an accuracy equivalent to one-hundredth of the power of a household light bulb is silly, especially if the method is only expected to provide an *estimate.* Three or four significant figures are satisfactory for a final result in most scientific or engineering studies. The important thing is to look at the accuracy of the method used and give an appropriate answer (i.e., engage the brain before transferring all nine digits from the calculator display into the report).

- Note the following points in regard to significant figures:
    - A stated result (measured or calculated) of 3.44 implies a true value between 3.435 and 3.445; similarly, a value of 3.4 implies a true value between 3.35 and 3.45.
    - The numerical result 3400 could imply a true value within the range 3350–3450 or a value in the range 3399.5–3400.5 (depending on whether the original value is understood to have two or four significant figures). This ambiguity can be eliminated by writing the number as either 3.4×10^2 or 3.400×10^2 (depending on the accuracy of the work).
    - The number of significant figures of a product or quotient is the minimum of the number of significant figures of the factors. For example, the volume (V) of a cylinder of radius r (where $r = 9.6$ mm) and height h (where $h = 14.3$ mm), should be written as:

    $$V = \pi r^2 h = 4.3 \times 10^3 \text{ mm}^2$$

    Note that the answer has two significant figures (i.e., not written as 4267 mm^2) as r has two significant figures, h has three and π has an infinite number of significant figures.

- When numbers are added or subtracted the least accurate component will determine the number of significant figures of the answer. For example, if $a = 45.3$ kg, $b = 17.8$ kg and $c = 6.52$ kg, the sum should be indicated as: $a + b + c = 69.6$ kg and not as 69.62 kg.

SI units

See **Units (SI)** for Système international d'unités (SI units).

Speculation

Speculation has its role in scientific work and it is perfectly acceptable to speculate on the probable reason(s) for your observations. Albert Einstein never conclusively proved his theory of relativity: his genius in creative thinking led him to postulate the theory, which at the time was supported by limited scientific evidence. The danger is to state as a matter of fact that which is only speculative. Do not conclude that something is true or false unless your argument is logically sound and the premises are correct (see **Logical conclusions**). There are many ways of framing an incorrect speculative argument—an extreme example: *This must be true because no other reason could be found.*

Spelling

- Your word processor has a spelling checker: always use it. When you are finished with the corrections, print the report and read it again. There is no excuse for submitting a report riddled with spelling errors, and do not believe for a minute that spelling is not important because you are writing in a scientific or technical discipline.
- Make sure that your word processor has the correct dictionary selected (e.g., American or British English). If the dictionary pro-

vides two acceptable spellings (e.g., *-ise* or *-ize*), be consistent in the use of the selected version.

- Do not change the name of an organization to comply with spelling rules. For example, the *British Centre for Strategic Research* should be spelled as indicated, and not with the American spelling of *Center*, even if the rest of the report has used standard American English spelling. Similarly, as **quotations** are exact transcripts of the original work, you should only use [*sic*] to indicate errors in the quotations and not correct variants of spelling.
- Foreign words should be spelled exactly as they are in their language, with the correct accents, cedillas, umlauts, etc. This does not apply to anglicized words (e.g., repertoire) that have been adopted into English with a change in spelling.
- See also **Incorrect word usage (confusables)**, **Hyphenation and word division**, and **Grammar and style [plurals]**.

Spreadsheets

Spreadsheets are great for analyzing engineering data, but if they are included as a figure in a report, without sufficient explanation, they can be exceptionally difficult, or even impossible, to comprehend. To give the readers a good chance to understand your results—and reproduce the spreadsheets if they wish—you need to define the equations in the cells and explain exactly what you have done. Furthermore, it may be helpful if the spreadsheets (with selected test cases) are recorded on a CD and submitted with the report.

Statistical analysis

- Statistical analysis can serve several purposes—for example:
  1. It can reduce a large amount of raw data to a manageable size, so that the readers will be able to quickly determine trends and patterns in the data.

(2) It can be used to determine if differences or correlations in the data are statistically significant or not.

- One of the main reasons for including details about the statistical analysis conducted is to give the readers confidence in the results presented. It also serves to provide information about the strength or weakness of a correlation.

- The two sets of numbers given in the table below have the same mean; yet a single glance down the columns will reveal that the two data sets are vastly different. This is blatantly obvious in this trivial illustration, yet many authors—even in reputable journals—fail to indicate what scatter existed in the data that they used to calculate the stated mean values. Worse still, some authors even fail to mention that a few data points were eliminated as being "outliers" from the rest of the data.

Table 3.1 Example of measured values

	Set A (mm)	Set B (mm)
Measurement 1	16.90	9.60
Measurement 2	16.50	31.60
Measurement 3	17.20	22.50
Measurement 4	16.50	16.20
Measurement 5	16.30	3.50
Mean	16.68	16.68

- Always indicate the number of samples in the data set and provide some way for the readers to assess the distribution of those values. For example, by including the standard deviation, or simply the maximum and minimum values, additional information will be provided that will be useful to the readers.

- See also **Error analysis (measurements)**.

Structure of reports and theses

Do not dream up a new report or thesis structure; reserve the imagination for the technical work and give the readers what they expect to see: a standard report. The structure of a typical report, with alternative headings, is given below, followed by the typical structure of a thesis.

See **Assessment report**, **Engineering report**, **Laboratory report**, and **Progress report** for further details on these report types. See also **Repetition** and **Reader**.

Typical structure of a report

Heading	Alternative heading	Notes
Title page		*Essential*
Summary	Abstract, Synopsis, Executive summary	*Essential*
Table of contents	Contents	*Essential*
List of figures	List of illustrations	*Optional*
List of tables		*Optional*
Nomenclature	Abbreviations, List of abbreviations, Symbols, List of symbols, List of abbreviations and acronyms, Glossary[a], Glossary of terms and abbreviations	*Essential if symbols or abbreviated terms are used*
1 Introduction	Background[b]	*Essential*
2 Objectives	Aims, Aims and objectives, Purpose, Scope	*Essential*
3 *You decide*	*You decide what you want to call the chapters in this part of the report. For a laboratory report, it could be*: 3 Theory, 4 Method, 5 Results; *and for an assessment report it could be*: 3 Tests conducted, 4 Results. *For other report types (e.g., engineering report or progress report), appropriate headings may be selected that best describe the work conducted.*	
4 *You decide*		
5 *Et cetera*		

Typical structure of a report (*continued*)

Heading	Alternative heading	Notes
6 Discussion		*Essential for most report types*
7 Conclusions	Concluding remarks, Summary of results, Conclusions and recommendations	*Essential*
8 Recommendations	Recommendations for further work, Suggestions for further study, Future research, Future work	*If required*
Acknowledgements[c]		*Optional*
References	List of references, Bibliography[d]	*Essential for most report types*
Appendix[e]	Annex, Addendum	*If required*

Notes:

[a] A *glossary* contains an explanation of terms, whereas a *nomenclature* defines the notation used. If extensive (i.e., more than about two pages), the *glossary* should be placed at the start or end of the *appendix* and, if appropriate, a *nomenclature* produced for the front of the report.

[b] A *background* chapter may follow the *introduction* if this element is not covered in the *introduction.*

[c] The *acknowledgements* are sometimes placed after the *summary*, which would be the traditional position in a thesis.

[d] A *bibliography* may contain works not cited in the report, but this is not the case for a *reference list.* Occasionally a writer may choose to have both a *reference list* and a *bibliography.*

[e] If there is more than one *appendix* the heading is *Appendices.*

Typical structure of a thesis

Heading	Alternative heading	Notes
Title page		*Essential*
Abstract	Summary	*Essential*
Acknowledgements		*Optional*
Table of contents	Contents	*Essential*
List of figures	List of illustrations	*Essential if there are a large number of figures*
List of tables		*Optional*
Nomenclature	Abbreviations, List of abbreviations, Symbols, List of symbols, List of abbreviations and acronyms, Glossary[a], Glossary of terms and abbreviations	*Essential if symbols or abbreviated terms are used*
1 Introduction[b]		*Essential*
2 Objectives[c]	Aims, Aims and objectives	*Essential*
3 *You decide*	*You decide what you want to call the chapters in this part of the thesis. The early chapters lay the foundation for what follows, and would frequently include a* Literature review[d] *followed by a* Theory *chapter. Subsequent chapters would deal with the experimental and theoretical work undertaken, each one starting with an introduction or overview of the chapter content and—if relevant—ending with chapter-related discussions and conclusions.*	
4 *You decide*		
5 *Et cetera*		
6 Discussion		*Essential*
7 Conclusions	Conclusions and recommendations	*Essential*
8 Recommendations	Recommendations for further research, Suggestions for further study, Future research	*If required*

Typical structure of a thesis (*continued*)

Heading	Alternative heading	Notes
References	List of references, Bibliography[e]	*Essential*
Appendix[f]		*If required*

Notes:

[a] A *glossary* contains an explanation of terms, whereas a *nomenclature* defines the notation used. If extensive (i.e., more than about two pages), the *glossary* should be placed at the start or end of the *appendix* and, if appropriate, a *nomenclature* produced for the front of the thesis.

[b] A *background* chapter may follow the *introduction* if this element is not covered in the *introduction*. Similarly, if the structure of the thesis is not described in the *introduction*, this too may follow as a separate chapter.

[c] A less used, alternative structure places the *objectives* before the *introduction*.

[d] Not a *literary review*, which has a different meaning.

[e] A *bibliography* may contain works not cited in the thesis, which is not the case for a reference list. Occasionally a writer may choose to have both a *reference list* and a *bibliography*.

[f] If there is more than one *appendix* the heading is *Appendices*.

Style (format)

See **Format style**.

Style manuals

- These are manuals for authors, editors, and publishers, detailing conventions for punctuation, spelling, abbreviations, capitalization, citation of references, nomenclature, word and numeral usage, and so forth. Many of the conventions are universally accepted as good practice and are, more or less, consistently described in the books—but there are also a lot of house rules peculiar to each style. These persnickety details are important to publishers, as they produce uniformity in style and format, but to casual writers of reports, they can be a source of confusion with contradictory recommendations galore. And there is no easy way to determine what is really important—to get the message across to the readers in an acceptable and unambiguous way—and what is only style preference.

- A list of well-known style manuals, serving a range of academic disciplines and publication formats, is given below.

  AIP Style Manual. 5th ed. Melville, NY: American Institute of Physics, 2000.

  Burchfield, R. W., ed. *The New Fowler's Modern English Usage.* Rev. 3rd ed. Oxford: Oxford University Press, 2000.

  Butcher, J., Drake, C., and Leach, M. *Butcher's Copy-editing: The Cambridge Handbook for Editors, Authors and Proofreaders.* 4th ed. Cambridge, UK: Cambridge University Press, 2006.

  Chicago Manual of Style: The Essential Guide for Writers, Editors, and Publishers. 15th ed. Chicago: University of Chicago Press, 2003.

  Coghill, A. M. and Garson, L. R., eds. *The ACS Style Guide: Effective Communication of Scientific Information.* 3rd ed. New York: Oxford University Press, 2006.

  Council of Science Editors. *Scientific Style and Format: The CSE Manual for Authors, Editors, and Publishers.* 7th ed. New York: Rockefeller University Press, 2006.

  Economist Style Guide. 9th ed. London: Profile Books, 2005.

  GPO Style Manual: An Official Guide to the Form and Style of Federal Government Printing. 30th ed. Washington: US Government Printing Office, 2008.

  Huth, E. J. *Medical Style and Format: An International Manual for Authors, Editors and Publishers.* Philadelphia: ISI Press, 1987.

  Iverson C., ed. *AMA Manual of Style: A Guide for Authors and Editors.* 10th ed. New York: Oxford University Press, 2007.

Merriam-Webster's Manual for Writers and Editors. Springfield, MA: Merriam-Webster, 1998.

Microsoft Manual of Style for Technical Publications. 3rd ed. Redmond, WA: Microsoft Press, 2004.

MLA Handbook for Writers of Research Papers. 7th ed. New York: Modern Language Association of America, 2009.

MLA Style Manual and Guide to Scholarly Publishing. 3rd ed. New York: Modern Language Association of America, 2008.

Publication Manual of the American Psychological Association. 6th ed. Washington, DC: American Psychological Association, 2010.

Ritter, R. *The Oxford Style Manual*. Oxford: Oxford University Press, 2003.

Rubens, P., ed. *Science and Technical Writing: A Manual of Style*. 2nd ed. New York: Routledge, 2001.

Siegal, A. M and Connolly, W. G. *The New York Times Manual of Style and Usage: The Official Style Guide Used by the Writers and Editors of the World's Most Authoritative Newspaper*. Rev. and expanded ed. New York: Times Books, 1999.

Turabian, K. L. *A Manual for Writers of Term Papers, Theses, and Dissertations*. 7th ed. Chicago: University of Chicago Press, 2007.

Style of writing (use of words)

- The different parts of the report are written using different prose styles. The **abstract** is technical and concise, with an economical use of words. The **introduction** has a flowing, natural style and reads as a story. The **method** resembles a recipe in a cookbook. The **results** are a dry, unadulterated record of the facts. The **discussion** allows greater opportunity for personal expression and flair. The **conclusions** contain a series of explicit statements—each carefully crafted to make a point—which are usually presented in a numbered or bulleted list.

- There are a few basic rules to follow when writing reports:

  (1) *Use serious, but normal, conversational language*: avoid long convoluted sentences and, if the context permits, use simple rather than fancy words (see below) and English rather than Latin (see **Latinisms**).

  (2) *Be precise*: avoid fuzzy, ambiguous or inexact statements (see below).

  (3) *Be concise*: get to the point promptly, do not labor the issue, and avoid wordy phrases and redundant text (see below).

  (4) *Be explicit*: avoid figures of speech (e.g., *open a can of worms*) and euphemisms (e.g., *the company must downsize*) (see below).

  (5) *Use formal language*: avoid colloquialisms, slang, inappropriate abbreviations, contractions, jargon, and exclamations (see below).

  (6) *Get the emphasis right*: structure ideas in sentences, lists, and paragraphs to convey, not just the information, but also the relative importance of the ideas (see below).

  (7) *Report results honestly and objectively*: do not read more into the results than is sensible, just because they are your results (see **Honesty** and also **Bad results**).

  (8) *Use bias-free terms*: avoid words that show bias with regard to gender, race, or culture.

Style of writing: (1) Fancy words

Fancy words—used to impress, rather than to communicate—add no value to a report and are likely to irritate readers. If a familiar, everyday word will do the job, use it.

Examples of fancy words

Fancy word	Possible synonyms
acquaint	advise, inform
adjourn	suspend, stop
affix	stick, attach
antithesis	opposite, converse
apprise	inform, tell
axiomatic	self-evident, unquestionable
bereft	stripped of, deprived of
capacious	roomy, spacious
conundrum	problem, difficult question
denigrate	belittle, diminish
desultory	casual, superficial
dubiety	uncertainty, doubtfulness
elucidate	explain, clarify
endeavor	try, attempt
impugn	challenge, question
maladroit	bungling, unskillful
modicum	little bit, small amount
peruse	read, study
post-haste	without delay, immediately
propensity	tendency, inclination
purvey	sell, supply
subterfuge	trickery, deception

Style of writing: (2) Examples of imprecise statements

Imprecise statement	Comment
The tests were reasonably successful.	Use numbers if you can—for example, rewrite the sentence as: *Sixty-five percent of the tests were successful.*
The plant is expected to be operational in the near future.	What is the *near* future: six weeks, eighteen months? Givc an indication of the time.

Style of writing: (2) Examples of imprecise statements (*continued*)

Imprecise statement	Comment
The gear failed after a high number of load cycles.	Provide an estimate if possible—for example: *The gear failed after a high number of load cycles (estimated to be of the order of 10^5).*
The revised test standard STN 2243-2 was consulted for the preparation of the specimens.	This implies that the test standard was *referred* to, but not that it was *adhered* to (an important distinction). If the latter is meant, rather say: *The specimens were prepared according to the revised test standard STN 2243-2.*
Failure to perform biweekly inspections will invalidate the insurance.	What is biweekly: twice a week or every two weeks? Avoid ambiguous terms.
Following the company's reorganization, there will be a need for less skilled technicians.	Does *less* refer to skill or technicians? Grammatically speaking, *less* should only be used for things that cannot be counted (*fewer* is used for things that can be); hence, it is *less skill*, not *fewer technicians*, that will be needed.
There were millions of data points that had to be examined.	Avoid the use of hyperbole (i.e., exaggerated statements, not meant to be taken literally).
The laboratory assistant removed the plastic model from the freezer after it had solidified and closed the door.	As inconceivable as it may be, this badly punctuated sentence attributes the closing of the freezer door to the plastic model (there should be a comma after *solidified*).
Real-time measurements were taken using a free stream ice crystal laser spectrometer.	It is not clear which adjectival noun applies to which term: is it a *crystal-laser spectrometer* that measures *free-stream ice* or a *laser spectrometer* that measures *free-stream ice-crystals*? Recast the sentence to eliminate ambiguity or use hyphens.

Style of writing: (3) Wordy phrases and redundant text

If you can cut out a word or two without affecting meaning, do so.

Examples of wordy phrases

Wordy phrase	Concise equivalent
as a matter of fact	actually, in fact
at the end of the day	eventually, finally
at this point in time	now
at this present moment in time	now
due to the fact that	because
for the purpose of (investigating)	to (investigate)
in order to	to
in spite of the fact that	though/although
in the event that	if
in the foreseeable future	later
in the long run	eventually
in view of the fact that	because
it is easier said than done	it is complicated
it stands to reason	it follows
it would appear that	apparently
last but not least	finally
on the occasion of	when
owing to the fact that	since
the reason why is that	because
they were of the same opinion	they agreed

Examples of redundant text

Redundant text	Concise equivalent
it was absolutely unique	it was unique
it was completely filled/emptied	it was filled/emptied
it was large in size	it was large
it was minimized to some extent	it was minimized
it was rectangular in shape	it was rectangular
it was somewhat maximized	it was maximized
it was yellow in color	it was yellow

Style of writing: (4) Figures of speech and euphemisms

- Figures of speech (e.g., idioms) and euphemisms can be puzzling to non-native English speakers; their use in technical writing is seldom justified (examples are given below). Clichéd phrases (e.g., a step in the right direction, all things being equal) should not be used.

Examples of figures of speech (to be avoided)

Figures of speech	Figures of speech
at the eleventh hour	*miss the boat*
bark up the wrong tree	*move up/down the food chain*
beat around the bush	*once in a blue moon*
bone of contention	*on the right track*
compare apples with oranges	*open a can of worms*
diamond in the rough	*par for the course*
draw a blank	*put two and two together*
fall between two stools	*sit on the fence*
in a nutshell	*start from scratch*
in the driving seat	*taken with a grain of salt*
it is a pipe dream	*the bottom line*
it is a white elephant	*the gold standard*
leave no stone unturned	*the Holy Grail*
level the playing field	*window of opportunity*

- The vast majority of *euphemisms* are substitutes for words connected with sex, excretion, religion, or death: topics not generally associated with technical writing. Nonetheless, there are a few euphemisms that are used by bureaucracies (e.g., government departments, military organizations) and corporations to express an objectionable action in a less offensive manner. This is occasionally called *doublespeak* and should be avoided—for example: *neutralize the target* (kill the enemy), *collateral damage* (unintentional killing of civilians), *right-sizing* or *downsizing* (firing of employees).
- Similarly, industrial organizations can take advantage of technical terminology to disguise something unpleasant—for example, *out-gassing* or *runoff* have been used as substitutes for *pollutants*.

Style of writing: (5) Formal writing

In a formal style of writing (which this guide does not use), you should avoid using

- Colloquial (i.e., not formal or literary) words and phrases—for example: *bummer*, *lab* (use *laboratory*), *lingo*, *math* (use *mathematics*), *savvy*, *stats* (use *statistics*), *tech* (use *technical*).
- Vulgarities and slang (e.g., the equipment was crappy; it arrived bloody late; the conclusions are bull; they sexed up the results).
- First-person pronouns (i.e., I, we, my, our, us); write in the **passive voice**, if necessary.
- Inappropriate abbreviations (e.g., the work @ Stanford; ½way; it was done B4; %age; OK; hydrogen & oxygen) and non-standard spelling (e.g., lite, slo, thru).
- Contractions—for example: *can't* (can not), *couldn't* (could not), *didn't* (did not), *don't* (do not), *shouldn't* (should not), *wouldn't* (would not), *it's* (it is or it has), *there's* (there is), *who's* (who is or who has), *they're* (they are), *you're* (you are).
- Jargon, if possible—frequently it is unavoidable. (Jargon denotes specialist words used by an expert group—for example: *myopia*, which means shortsightedness.)
- Rhetorical questions (e.g., Why are these results important?).
- Exclamations (e.g., These results are 50% better than those reported by Yang!).
- Incomplete sentences (see **Grammar [incomplete sentences]**).
- Gimmicks, for example: smileys :-) and smiley icons ☺, banners (around headings, for example), clip art (when it serves no purpose), and non-standard bullets (e.g., ♥, ♫) —see also **Gimmicks, clip art, and emoticons (smileys)**.

Style of writing: (6) Emphasis

- Emphasis is conveyed by selecting the right words (which is obvious) and the order in which the ideas are presented (which is also obvious, but frequently messed up). Creating suspense by making readers wait—for the punch line or the closing remark that makes everything clear—is gripping stuff in a novel, but reports are not for leisure!
- As a general rule, you should get the most important information across first, followed by secondary details or explanatory elements (which clarify, justify, or limit the main point). This approach can be used to structure observations, concepts, conclusions, and so forth in paragraphs, lists, and individual sentences.

  Example:

  > Extensive corrosion and internal cracking resulted in the aluminum test pieces failing prematurely and, as there was no observation of degradation of the material during eighteen months of environmental testing, the optimum material for the manufacture of Dumpty's replacement shell was confirmed to be titanium.

  The writer has three things to say, but we have to wait to the end to get the main point. The sentence can be improved by turning it around:

  > Titanium has been confirmed as the optimum material for the manufacture of Dumpty's replacement shell (*main point*), as no degradation of the material was observed during eighteen months of environmental testing (*justification*), whereas the aluminum test pieces failed prematurely due to extensive corrosion and internal cracking (*secondary point*).

- See also **Passive voice**, **Tenses (past, present, and future)**, and **Grammar and style**.

Summary

- A summary (sometimes called a *synopsis*) is an overview of the work undertaken. The terms *summary* and *abstract* are often used interchangeably; however, there is a subtle distinction: a summary would be written for a less technical readership, using language more easily understood by the general public than would be the case for an abstract. In all other respects the requirements for a summary would be the same as those for an abstract.
- It is not common for a report to have both a summary and an abstract. In the rare situation where this is required, it is probably because the summary will be stored separately and made available to a wider, non-technical audience. The summary would in that case be similar to the abstract, but with an extended description of the results and conclusions using less technical language.
- See also **Abstract** and **Executive summary**.

Summary of results

This heading may be used instead of **conclusions** when the author does not wish to make definite conclusions deduced from the results, but simply wishes to highlight the important results without comment.

See also **Conclusions** and **Concluding remarks**.

T

Table of contents

- The table of contents appears on its own page under the heading *contents* or *table of contents*. It serves as an inventory for the report, identifying all major parts and indicating the corresponding page numbers.
- The heading structure for large reports may comprise several levels (see **Headings**), but it is not necessary for the lower levels to be represented in the table of contents: two or three levels are usually adequate.
- Chapters, sections and sub-sections should be numbered using Arabic numerals. Note that the *reference list* and the *appendix* are not considered to be chapters in the conventional sense and are usually not allocated chapter numbers.
- The automatic generation of a table of contents using a word processor saves a lot of time, but it requires forward planning and a consistent use of a **format style** for each heading level.
- Many reports are not *read* from cover to cover; instead they are *used* to find interesting and relevant pieces of information, with the reader flicking repeatedly from the table of contents to the text. So do not forget to check the table of contents and update the page numbers before submitting the report; it is an important navigation tool, and if it is incorrect, it will frustrate your readers immensely.
- For a large report, it is advisable to provide a *list of figures* and a *list of tables* to supplement the table of contents (anything that makes it easier for readers to find their way around the report is a good idea).

Example of table of contents for an engineering report

TABLE OF CONTENTS

Tables

Tables—as distinct from **figures**—have rows and columns of data (flowcharts and schematics with blocks of text are treated as figures). Data presentation is a critical element of most engineering reports, so it is worth paying attention to the conventions for formatting tables.

Structure of a table (typical)

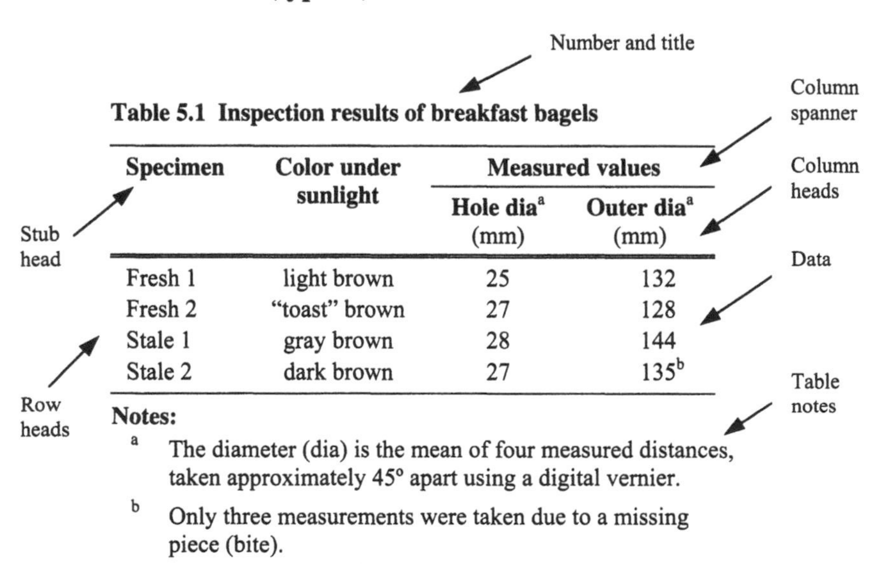

Table 5.1 Inspection results of breakfast bagels

Specimen	Color under sunlight	Measured values	
		Hole dia[a] (mm)	Outer dia[a] (mm)
Fresh 1	light brown	25	132
Fresh 2	"toast" brown	27	128
Stale 1	gray brown	28	144
Stale 2	dark brown	27	135[b]

Notes:

[a] The diameter (dia) is the mean of four measured distances, taken approximately 45° apart using a digital vernier.

[b] Only three measurements were taken due to a missing piece (bite).

Tables: number and title

- Tables should be numbered using Arabic numerals in the order in which they are *first* mentioned in the report (even if a more detailed discussion appears later). For a long report, it is convenient to use a two-part number system: the first identifying the chapter and the second allocated sequentially within each chapter (e.g., *Table 5.1* will be the first table in Chapter 5).
- Each table should have a title, which is placed *above* the table (this convention is different from that used for figures, where the caption is located *below* the figure). Titles are often written in sentence style (i.e., starting with a capital letter, followed by all words in lowercase,

except for proper nouns and certain acronyms and abbreviations), but that is only a matter of style preference. A period is not needed after the title. The selected style should be consistently applied throughout the report.

Tables: column heads (headings) and spanner

- Each column should have a short, descriptive heading (no period is needed after the heading). Use abbreviations to condense the table. Standard abbreviations (e.g., *No.* for *number*) do not require an explanation, but non-standard abbreviations should be defined using table notes (tables should be intelligible without reference to the text).
- Use a column spanner (i.e., a heading that spans several columns) to group related table entries and eliminate the replication of column heading information.
- Headings should ideally be set horizontally. However, if the headings are long and do not fit across the page, the text may be set at an angle (say 45°) or turned through 90°. The convention is that the page be rotated clockwise to read the text—for example:

Bread									
French baguette	Greek kouloura	Indian chapatti	Indian naan	Irish soda	Lebanese pitta	Mexican tortilla	N. American corn	Russian black	Scandinavian rye

- Information pertaining to all entries in a column should be written as a subheading or as a table note, rather than after each entry.
- Units are best placed below the heading—centralized and in parentheses—rather than following table entries. This, obviously, only works if the units apply to all entries in a column.

- Occasionally it is convenient to number the columns in order to make reference to table entries in the text. In such cases it is best to number all columns, starting from the left and to set the column numbers in parentheses.

Tables: row heads (headings)

- Row headings, like column headings, should be short and descriptive. Secondary details should rather be included as table notes than squashed into a little box (or cell).
- It is better to place subheadings (i.e., categories within each row heading) below the row headings (indented or right justified) than in the adjacent column—for example:

Good option

Grape variety	**Sugar content**
Red	
Gamay	data
Merlot	data
Shiraz	data
White	
Chardonnay	data
Riesling	data
Semillon	data

Poor option

Grape variety		**Sugar content**
Red		
	Gamay	data
	Merlot	data
	Shiraz	data
White		
	Chardonnay	data
	Riesling	data
	Semillon	data

Tables: stub head (a heading for the row heads)

- Use a stub heading, which is a heading for the row headings (e.g., *Grape variety* in the above example), whenever possible.
- Do not split the stub head box (cell) to cater for both row and column headings. It is best to use the stub head box for row headings only and to add a column spanner for the column headings.

Good option

Cheese	Feature		
	Head 1	**Head 2**	**Head 3**
Brie	data	data	data
Cheddar	data	data	data
Gouda	data	data	data
Parmesan	data	data	data

Poor option

Feature \ Cheese	Head 1	Head 2	Head 3
Brie	data	data	data
Cheddar	data	data	data
Gouda	data	data	data
Parmesan	data	data	data

Data in tables

- As it is easier to compare numbers in columns than in rows (vertical scanning is easier than horizontal scanning), structure the table with *columns* of similar data.
- Numbers should be expressed as numerals, never words.
- Leave the cell blank if data could not be obtained or is not applicable (it may be helpful to the reader to explain in a footnote why the cell is empty). However, a numerical value of zero should be entered as 0 (the cell should not be left blank in this case).
- Never use *ditto* or *do* to indicate that a number is repeated; write out the number each time.
- Numbers within columns should be aligned about the decimal point, or in the absence of decimal points, justified to the right (i.e., with the units, tens, and hundreds aligned).
- Numerically large numbers in tables can be difficult to handle. For example, if you wish to record the following three values in a column: 1 344 000 kg, 1 315 000 kg, and 1 382 000 kg, then it is acceptable to write the numbers in the table as 1.344, 1.315, and 1.382, and to write the units in the column heading as 10^6 kg. Do not write the units as kg 10^{-6}, arguing that the numbers have been divided by one million, as this is incorrect.

- Units, symbols, currencies, and so forth that apply to all items in a column are best located below the column header (within parentheses) and not within the body (i.e., data field) of the table.
- Select data for the table carefully: cut out non-essential information and omit columns that can be easily calculated from other columns.

Table notes (table footnotes)

- Table notes (table footnotes) are placed immediately below the table, and can be used, for example, to define an abbreviation or to provide an explanation of a particular entry or procedure adopted. Table notes should be numbered separately and should not be part of the numbering system used for footnotes that appear in the text. Use lowercase letters (a, b, c, etc.) or symbols (* † ‡ § ¶) to make this distinction if necessary. The sequence begins afresh for each table.
- If you are not personally responsible for the data in the table, you should credit the original author(s). This is usually done by writing "Source:" immediately below the table, followed by the details of the original work. If only part of the table comes from a particular source, make this clear using a footnote or alternatively add another column to the right of the table to indicate the source of the data that was recorded in that particular row. Use phrases such as *Adapted from …* or *Based on …* if you modified or re-formatted the data (see also **Copyright**, **Fair-use doctrine**, and **Plagiarism**).

Tables: grids and field spanners

- The inclusion of grid lines in tables is a matter of personal taste. In most cases, vertical grids are superfluous. Publishers tend to use as few grids as possible, while ensuring that the table is still intelligible.
- Field spanners span the columns within the data field, providing areas to identify a third data variable. This table structure is only necessary when the row headings already have two variables defined and a third is needed. In the following example, the third variable is the "region." The alternative, of course, is to have two tables.

Grape variety	Head 1	Head 2	Head 3	Head 4
	California			
Red				
Merlot	data	data	data	data
Zinfandel	data	data	data	data
White				
Chardonnay	data	data	data	data
Chenin blanc	data	data	data	data
	France			
Red				
Cabernet sauvignon	data	data	data	data
Merlot	data	data	data	data
White				
Chardonnay	data	data	data	data
Semillon	data	data	data	data

Integration of tables into the report

- All tables should be referred to in the text (see **Cross-referencing**) and should be located in the report soon after their first mention. Draw the reader's attention to the table and highlight the most important elements of the table (if you discuss every detail, the table will be redundant).
- Try to make the table self-explanatory and self-contained, with all parts of the table (including footnotes) appearing on a single page. If the table has to be split, indicate this on the subsequent pages, by writing *Table 5.1 continued*, for example, and assist the readers to understand the table by repeating the column headings.
- If you have a lot of tables in the report, prepare a list—which should be placed after the *table of contents* (see **List of figures/tables**). The facility in word-processing software to automatically generate such lists, complete with page numbers, is extremely useful; however, it requires planning and the selection of a single **format style** for the titles.

Technical dictionaries

Like ice cream, technical dictionaries come in all conceivable flavors: from vanilla to strawberry cheesecake. A sample:

ASTM Dictionary of Engineering Science & Technology. 10th ed. West Conshohocken, PA: ASTM International, 2005.

Bignami, M. *Elsevier's Dictionary of Engineering*. In English, German, French, Italian, Spanish and Portuguese. Amsterdam: Elsevier, 2004.

Blockley, D. *The New Penguin Dictionary of Civil Engineering*. Harmondsworth, UK: Penguin, 2005.

Chakolov, G. *Elsevier's Dictionary of Science and Technology*. In Russian–English (1993), English–Russian (1996). Amsterdam: Elsevier.

Dorian, A. F. *Dorian's Dictionary of Science and Technology*. In English–German (1989), German–English (1981), English–French (1979), French–English (1980). Netherlands: Elsevier.

Gattiker, U. E. *The Information Security Dictionary*. Boston: Kluwer Academic Pub., 2004.

Gosling, P. J. *Dictionary of Biomedical Sciences*. London: CRC Press, 2002.

Hargrave, F. *Hargrave's Communications Dictionary*. New York: IEEE Press, 2001.

Harris, C. M., ed. *Dictionary of Architecture & Construction*. 4th ed. New York: McGraw-Hill, 2006.

IEEE 100: The Authoritative Dictionary of IEEE Standards Terms. 7th ed. New York: IEEE Press, 2000.

IFIS Dictionary of Food Science and Technology. 2nd ed. Chichester, UK: Wiley-Blackwell, 2009.

Lackie, J. *Chambers Dictionary of Science and Technology*. New rev. and updated ed. Edinburgh: Chambers, 2007.

Laffery P. and Rowe, J., eds. *The Hutchinson Dictionary of Science*. Oxford: Helicon, 1994.

Laplante, P., ed. *Dictionary of Computer Science, Engineering, and Technology*. London: CRC Press, 2001.

McGraw-Hill Dictionary of Bioscience. 2nd ed. New York: McGraw-Hill, 2003.

McGraw-Hill Dictionary of Chemistry. 2nd ed. New York: McGraw-Hill, 2003.

McGraw-Hill Dictionary of Engineering. 2nd ed. New York: McGraw-Hill, 2003.

McGraw-Hill Dictionary of Environmental Science. New York: McGraw-Hill, 2003.

McGraw-Hill Dictionary of Mathematics. 2nd ed. New York: McGraw-Hill, 2003.

McGraw-Hill Dictionary of Physics. 3rd ed. New York: McGraw-Hill, 2003.

McGraw-Hill Dictionary of Scientific and Technical Terms. 6th ed. New York: McGraw-Hill, 2003.

Morris, C., ed. *Academic Press Dictionary of Science and Technology*. UK: Academic Press, 1996. [CD].

Porteous, A. *Dictionary of Environmental Science and Technology*. 4th ed. New York: Wiley, 2008.

Schramm, L. L. *Dictionary of Colloid and Interface Science*. 2nd ed. New York: John Wiley, 2001.

Walker, P. M. B., ed. *Chambers Materials Science and Technology Dictionary*. Edinburgh: Chambers, 1993.

Weik, M. *Computer Science and Communications Dictionary*. Boston: Kluwer Academic Pub., 2000.

Tenses (past, present, and future)

- Do you write in the *past*, *present*, or *future tense*? This is not always an easy question to answer. For the most part the report is written in the past tense, but there are many situations where it is more appropriate to write in the present tense, and a few situations—usually restricted to the recommendations—when the future tense is used.

- Use the past tense to describe
  (1) The work undertaken or process adopted (e.g., The experiment *was* performed; Invalid data points *were* eliminated);
  (2) The behavior of models, test specimens, and so forth and the results obtained (e.g., There *was* little evidence of corrosion);
  (3) The results of others (e.g., Khan[7] *measured* sulfur levels); or
  (4) The equipment or apparatus used (permanent facilities, however, are usually described in the present tense).

- But use the present tense
  (1) To draw the reader's attention to something (e.g., The results *are* summarized in Table 5.2; Figure 4.1 *indicates* ...);
  (2) To describe accepted/established facts or existing situations or conditions (e.g., The most important greenhouse gas *is* CO_2; The bridges *are* not at risk of collapsing);
  (3) To make an observation or conclusion or to explain something (e.g., The results in Table 5.3 *show* ...; The analysis *suggests* that ...; This result *is* attributed to ...); or
  (4) When reporting on the opinion of others relating to a current topic (e.g., Lopes(56) *maintains* that the role of cirrus in the radiative heating of the Earth *is* poorly understood).

Theory

- If the theory that needs to be reviewed or developed is extensive, this will justify a chapter in its own right; otherwise it can be included as part of the **introduction**. If the theory is likely to be of only peripheral interest to the reader, it is best placed in the **appendix**.
- Be meticulous in distinguishing between that what is already known (and reported elsewhere) and what you are presenting as original work.

- Deciding where to start developing the theory is a tricky issue: if you begin at an elementary level, you will bore the readers (wasting valuable time—yours and theirs) and if you only provide the final few lines of a complex mathematical derivation, it will not be widely understood. So you need to start somewhere between these extremes and establish a foundation: a series of starting points that can be referenced to well-established textbooks or other reputable publications (e.g., journal or conference papers).
- Develop the theory at a pace that can be followed by someone who has a good general understanding of the subject area, but is not an expert in the specific field. A helpful technique is to envisage yourself explaining the theory to someone who has a technical knowledge comparable to that which you had before you started the particular study that you are now describing.
- See also **Mathematical notation and equations**.

Title of report

Every report must have a title; it should be descriptive, but not too long (i.e., not more than 100 characters). Use a sub-heading if necessary.

See also **Front cover** and **Title page**.

Title page

- Bound formal reports and theses usually have a title page, which will provide the same information as the **front cover** and a few additional facts. For less formal reports, a title page is not necessary when all relevant information is provided on the front cover.
- The title page of a *report* would typically indicate
  - Title of the report;
  - Volume number and the total number of volumes (if appropriate);
  - Report number (if appropriate);

   - Names of the authors and their qualifications (optional);
   - Name of the organization and address (if appropriate); and
   - Date.

- The title page of a *report* could also include
   - Name(s) and affiliation(s) of the supervisor(s) of the work (if appropriate);
   - Total number of pages;
   - Restriction on distribution, say for reasons of confidentiality;
   - Copyright statement;
   - The reason for writing the report and/or its recipient (e.g., Report submitted to the Bureau of Accident Investigation under contract number 2004/019); or
   - Declaration of originality (e.g., I/We declare that this is my/our work and that all contributions from other persons have been appropriately identified and acknowledged).

- A typical *thesis* title page will indicate
   - Name of the university/institute and, optionally, the name of the school/college/faculty;
   - Degree award;
   - Academic years;
   - Name of candidate;
   - Title of thesis;
   - Name(s) and affiliation(s) of the supervisor(s); and
   - Date.

- The title page of a *thesis* could also include
   - Statement of submission (e.g., A thesis submitted in partial fulfillment of the requirements for the degree of Bachelor of Science in Viticulture); or
   - Declaration of originality (e.g., I declare that this is my own work and that it has not been previously submitted to this or any other academic institution for this or any other academic award).

Trade names and trademarks

The use of *trade names* is generally discouraged in academic and scientific organizations if the report is going to enter widespread distribution, as these organizations generally do not wish to endorse or advertise commercial products. Generic names should be used whenever possible—for example: use *petroleum jelly* rather than *Vaseline*. Other trade names adopted into general use that you should avoid are *Band-Aid*, *Hoover*, *Photostat*, *Sellotape*, *Teflon* and *Xerox* (note that they are capitalized) However, if the test results of your particular study are dependent on a specific supplier's material being used, or if replication of the test requires the use of a specific piece of equipment, then it is appropriate to indicate the trade name of the product used.

A *trademark* is a sign (comprising words, letters, numerals, logos, pictures, or a combination of these) that is used to distinguish the products or services of one organization from those of another. The trademark symbol ® indicates that the trademark has been registered, whereas the symbol ™ indicates that the supplier wishes the trademark to be recognized as a distinguishing sign of the organization's products or services (say pending registration). Companies often insist that their trademark name be written in a particular way (e.g., in capital letters, such as IMAX, or with certain phrasing, such as Windows NT). There is no legal requirement to use the trademark symbols ® and ™ following the name of a company or product, but if they are used, it may be helpful to the readers if you give the company name and shortened address as a footnote when first mentioned.

Turabian

Kate Turabian wrote a popular little book in 1937 called *A Manual for Writers of Term Papers, Theses, and Dissertations*. Now in its seventh edition (University of Chicago Press, 2007), it remains a widely consulted work, particularly by students who balk at the cost of the *Chicago Manual of Style* and want lots of examples on how to cite references.

See also **Style manuals**.

Types of technical and scientific documents

conference paper An essay that accompanies a conference presentation, of varying format: it may follow the basic structure of a report, but without table of contents or lists of figures and tables, and typically without appendices. It is usual that only the abstract or summary (and not the whole paper) is peer reviewed (i.e., approved by domain experts) prior to acceptance.

dissertation From the Latin for *discussion*, usually synonymous with thesis (see below), but occasionally used exclusively for doctoral work.

journal paper/note A concise and narrowly focused essay describing original work. It follows the basic structure of a report, but without table of contents, lists of figures and tables, and usually without appendices. Journal entries of reputable journals are peer reviewed (i.e., approved by domain experts) prior to publication and must conform to the publisher's house rules regarding format.

monograph A detailed written study (usually a book-length work) of a single specialized subject (e.g., the role of cirrus in the radiative heating and cooling of the Earth).

paper A long essay on one or more subjects, not necessarily complying with the traditional report structure, but written in a formal manner that would include the appropriate citation of references.

report In this context, a report is a structured written account, or record, of a study, investigation, experiment or analysis. Examples:

(1) assessment report A written record following the assessment of the capabilities or condition of an artifact or individual (e.g., the condition of a building or a patient's heart), detailing the history or background, assessment procedure, findings and recommendations.

(2) engineering report A written record of an engineering study or investigation (e.g., concerning the heat generated by a mobile phone), describing the theoretical and/or experimental work conducted, results and conclusions—often accompanied by appendices containing supporting information of sufficient detail that would enable a peer to validate the work.

(3) **laboratory report** A written record of a laboratory investigation or experiment (e.g., crystal growth in monoammonium phosphate solutions), describing the procedure, materials, and results in sufficient detail to enable a peer to replicate the work.

(4) **progress report** A written record of work completed, usually relating to a project, describing, for example, the tasks undertaken, time schedule, milestones reached, delays encountered, deliverables submitted and costs incurred; usually submitted to a funding agency or external organization.

thesis A long, structured essay describing personal research, usually written by a candidate as part or whole submission for consideration by a university for the award of a degree (e.g., M.Sc. thesis, doctoral thesis). It would comply with the traditional report structure, with some minor variations, including a cover (in addition to a title page), a literature review section/chapter and frequently a theory section/chapter. The acknowledgements would usually follow the abstract, and would not be placed at the end of the document, as is typical for a report.

U

Units

- Units of measurement are vital information: never omit them!
- The use of SI units for engineering work in the US (particularly in organizations with international subcontractors or subsidiaries) has increased significantly in recent years; it is also the preferred option for most engineering publishers. As a general rule, you should avoid switching between SI and non-SI units—if you need to have both millimeters and inches in the report, for example, choose one system as the primary measure and consistently give the equivalent values in parentheses afterwards.
- In the case of measurements of physical properties (e.g., the weight of a sample) or specifications for parts manufacture (e.g., the positions of holes in a metal coupling), it is best to report the measurements or dimensions in their original units (and to provide the equivalent, converted, values in parentheses).
- Include a table of appropriate conversion factors in the **appendix** if you think it will be helpful to the readers.
- Be consistent in the selection of units for presenting your results—graphs and tables displaying the same or similar type of data should have the same units to facilitate a direct comparison to be made (e.g., do not use cm for one graph and mm for another).

Format for writing numerals and their units

- Most **style manuals** recommend putting a single space between the numeral and the unit (e.g., 17.3 kN) rather than writing them "closed up" (e.g., 50Hz). There is one exception: it is customary not to put a space between a numeral and a superscript-type unit symbol (e.g., 60°, 45" 15'). This applies to the degree symbol when used as a measure of plane angle; however, when used as a measure of temperature, many authorities (including the International Organization for Standardization) recommend that the space be retained (e.g., 15 °C, 45 °F). Nevertheless, these are matters of style preference; you should select a style for your report and use it consistently.

- Do not split a numeral and its unit at the end of a line. (Use a "non-breaking space"—obtained in Microsoft Word by clicking on the Insert pulldown menu and selecting Symbol [Special characters], or alternatively by using the shortcut keys Control+Shift+Space.)

- Use roman (upright) and not italic typeface for unit symbols (unit abbreviations), which should be written without periods (except, of course, when the unit falls at the end of the sentence).

- Use the same symbol (abbreviation) for both singular and plural (i.e., never add an *s* to a unit symbol; the following are incorrect: 2 kgs, 3 lbs, 4 mins).

- Use a raised dot or a space to separate compound units (e.g., ft·lbf or ft lbf).

- Do not write compound units that have a product of terms in the denominator with more than one slash (e.g., avoid mg/N/s or lb/lb/h). Such formulations can be confusing to readers not familiar with the subject. Use the format mg N^{-1} s^{-1} or mg·N^{-1}·s^{-1} (with a raised dot) when the denominator has more than one term.

- Always use the symbol (abbreviation) and not the name of the unit following a numeric value (e.g., write 12.2 V and not 12.2 volts); however, when the unit is mentioned without an accompanying value the name may be used (e.g., Temperature measurements were recorded in degrees Celsius).

- When a measurement (or quantity) is used as a modifier before a noun, there should not be a hyphen (or dash) between the numeral and the unit symbol (abbreviation)—for example, do not write: *It was a 0.5-kg block.* If, however, the unit is spelled out—which is generally not the case in engineering reports—then a hyphen is traditionally used (e.g., It was a half-kilogram block).

- See also **Units (conversion factors)**, **Units (foot-pound-second, CGS, and miscellaneous)**, and **Units (SI)**.

Units (conversion factors)

When converting terms (say from standard English units to SI units) ensure that the new (i.e., converted) values do not imply a level of precision or accuracy that is inconsistent with the original values—for example, do not translate a *1 lb hammer* as a *0.4536 kg hammer* or write *9 MPa* when the original was *1.32 ksi*. A practical guideline, used by many engineers, is to write the new value with one additional significant figure—for example: *81 hp (60.4 kW)*.

Conversion factors (to SI units)

Unit	SI unit
length	
1 in	= 25.4 mm (exactly)
1 ft	= 12 in (exactly) = 0.304 8 m (exactly)
1 yd	= 3 ft (exactly) = 0.914 4 m (exactly)
1 mile	= 5 280 ft (exactly) = 1 609.344 m (exactly)
1 nm	= 1 852 m (exactly)
area	
1 in^2	= 645.16 mm^2 (exactly)
1 ft^2	= 0.092 903 04 m^2 (exactly)
1 yd^2	= 0.836 127 36 m^2 (exactly)
1 $mile^2$	= 2.589 988 km^2
1 acre	= 4 046.856 m^2
volume	
1 in^3	= 16.387 064 cm^3 (exactly)
1 ft^3	= 28.316 85 dm^3 (exactly)
1 yd^3	= 0.764 554 9 m^3
1 fl oz (UK)	= 28.413 06 cm^3
1 fl oz (US)	= 29.573 53 cm^3
1 pt (UK)	= 0.568 261 25 dm^3 (exactly)
1 liq pt (US)	= 0.473 176 5 dm^3
1 dry pt (US)	= 0.550 610 5 dm^3
1 gal (UK)	= 277.420 in^3 = 4.546 092 dm^3 (exactly)
1 gal (US)	= 231 in^3 = 3.785 412 dm^3
1 barrel	= 9 702 in^3 = 158.987 3 dm^3

Conversion factors (to SI units) (*continued*)

Unit	SI unit
1 bbl (dry, US)	= 7 056 in^3 = 115.627 1 dm^3
1 L	= 1 dm^3 (exactly)
velocity	
1 ft/s	= 0.304 8 m/s (exactly)
1 mile/h	= 0.447 04 m/s (exactly)
1 km/h	= (1/3.6) m/s (exactly) = 0.277 778 m/s
1 kn	= 0.514 444 m/s
acceleration	
1 ft/s^2	= 0.304 8 m/s^2 (exactly)
mass	
1 lb	= 0.453 592 37 kg (exactly)
1 oz	= (1/16) lb = 28.349 52 g
1 ton (UK)	= 2 240 lb (exactly) = 1 016.047 kg
1 long ton	= 2 240 lb (exactly) = 1 016.047 kg
1 ton (US)	= 2 000 lb (exactly) = 907.184 7 kg
1 t	= 1 000 kg (exactly) (called tonne or metric ton)
density	
1 lb/ft^3	= 16.018 46 kg/m^3
force	
1 lbf	= 4.448 222 N (based on the standard value of gravitational acceleration)
1 kgf	= 9.806 65 N (based on the standard value of gravitational acceleration)
moment of force	
1ft·lbf	= 1.355 818 N·m
pressure or *stress*	
1 lbf/in^2	= 6 894.757 Pa
1 atm	= 101 325 Pa (exactly)
1 in Hg	= 25.4 mm Hg = 3 386.389 Pa
1 Torr	= 133.322 4 Pa

Conversion factors (to SI units) (*continued*)

Unit	SI unit
angle (plane)	
1°	= (π /180) rad = 0.017 453 3 rad
second moment of area	
1 in^4	= 41.623 14 x 10^{-8} m^4
section modulus	
1 in^3	= 16.387 064 x 10^{-6} m^3 (exactly)
kinematic viscosity	
1 ft^2/s	= 0.092 903 04 m^2/s
energy	
1 ft·lbf	= 1.355 818 J
power	
1 ft·lbf/s	= 1.355 818 W
1 hp	= 550 ft·lbf/s (exactly) = 745.699 9 W
temperature	
1 °R	= (5/9) K
°F to °C	$t_C = (5/9)(t_F - 32)$ (where t_C and t_F are temperature values in °C and °F respectively)
°C to K	$t_K = t_C + 273.15$ (where t_K and t_C are temperature values in K and °C respectively)
heat	
1 Btu	= 788.169 ft·lbf = 1 055.056 J
1 Btu/h	= 0.293 071 1 W
1 Btu/(s·ft·°R)	= 6 230.64 W/(m·K)
1 Btu/(h·ft^2·°R)	= 5.678 26 W/(m^2·K)
1 Btu/(lb·°R)	= 4 186.8 J/(kg·K) (exactly)

Notes:

1 Source: ISO 31:1992 (Specification for quantities, units and symbols).

2 See also **Units (foot-pound-second, CGS, and miscellaneous)**.

Units (foot-pound-second, CGS, and miscellaneous)

Units: foot-pound-second (fps), CGS[1], and miscellaneous

Unit	Symbol or abbreviation	Notes
acidity *or* alkalinity	pH	measure of acidity *or* alkalinity
acre	acre	no abbreviation
angstrom	Å	$= 10^{-10}$ m
atmosphere (standard)	atm	
bar	bar	= 100 kPa
barrel	barrel	$= 9\ 702$ in^3 (for petroleum, etc.)
barrel, dry (US)	bbl	$= 7\ 056$ in^3
baud	baud	no abbreviation
bit per second	bps	
brake horsepower	bhp *or* BHP	
British thermal unit	Btu	
bushel (UK)	bushel	= 8 gal (UK)
bushel (US)	bu	= 64 dry pt (US)
cubic foot	ft^3 *or* cu ft	use ft^3 rather than cu ft
cubic inch	in^3 *or* cu in	use in^3 rather than cu in
day	d	spell out if ambiguous
degree	°	see note 2
degree Celsius	°C	
degree Fahrenheit	°F	
degree Rankine	°R	
dots per inch	dpi	
calorie	cal	alternative definitions are used
decibel	dB	
dyne	dyn	$= 10^{-5}$ N (CGS unit)
erg	erg	$= 10^{-7}$ J (CGS unit)
fluid ounce (UK)	fl oz	160 fl oz (UK) = 1 gal (UK)
fluid ounce (US)	fl oz	128 fl oz (US) = 1 gal (US)
foot	ft *or* ′	use ft rather than ′
foot-pound	ft·lb *or* ft·lbf	
gallon (UK)	gal	1 gal (UK) = 277.42 in^3
gallon (US)	gal	1 gal (US) = 231 in^3
gigabit	Gbit	avoid abbreviation; spell out
gigabyte	GB	$= 2^{30}$ bytes = 1 073 741 824 bytes

Units: foot-pound-second (fps), CGS, and miscellaneous (*continued*)

Unit	Symbol or abbreviation	Notes
grain	gr	= (1/7000) lb
hectare	ha	= 10^4 m^2
horsepower	hp	
hour	h *or* hr	use h rather than hr
hundredweight	cwt	= 112 lb
inch	in *or* ″	use in rather than ″
inch-pound	in·lb *or* in·lbf	
inch mercury	in Hg	
kilobit	Kb	avoid abbreviation; spell out
kilobit per second	Kbps	
kilobyte	KB	= 2^{10} bytes = 1 024 bytes
kilobyte per second	KBps	
kilogram-force	kgf	rather use SI unit
kilopond	kp	= kgf (not recommended)
kilopound	kip	= 1 000 lb
knot	kn *or* kt	nautical mile per hour
liter	L *or* l	see note 3
megabit	Mb	spell out if ambiguous
megabit per second	Mbps	
megabyte	MB	= 2^{20} bytes = 1 048 576 bytes
megabyte per second	MBps	
micron	µm *or* µ	µm complies with SI
mile	mi	spell out if ambiguous
mile per gallon	mi/gal *or* mpg	
mile per hour	mi/h *or* mph	
milliseconds	ms *or* msec	ms complies with SI
minute	min *or* ′	avoid ′ for time
mole	mol	amount of substance
nautical mile	nm *or* naut mi	
ounce	oz	
parts per million	ppm	
pint (UK)	pt	8 pt (UK) = 1 gal (UK)
pint, dry (US)	dry pt	64 dry pt (US) = 1 bu (US)
pint, liquid (US)	liq pt *or* pt	8 liq pt (US) = 1 gal (US)
point	pt	
point per inch	ppi	

Units: foot-pound-second (fps), CGS, and miscellaneous (*continued*)

Unit	Symbol or abbreviation	Notes
poise	P	= 10^{-1} Pa·s (CGS unit)
pound	lb	fps unit of force is lb *or* lbf
pound-force	lbf	
pound per square inch	psi	= lbf/in^2
revolution per minute	r/min or rpm	
second	s *or* sec *or* ″	s complies with SI; avoid ″ for time
slug	slug	no abbreviation; fps unit of mass
square foot	ft^2 *or* sq ft	use ft^2 rather than sq ft
square inch	in^2 *or* sq in	use in^2 rather than sq in
stokes	St	= 10^{-4} m^2/s (CGS unit)
ton (UK) *or* long ton	ton	= 2 240 lb (long ton used in US)
ton (US)	ton	= 2 000 lb
tonne *or* metric ton	t *or* tn	= 1 000 kg (metric ton used in US)
torr	Torr	rather use SI unit
week	wk	spell out if ambiguous
yard	yd	
year	y *or* yr	spell out if ambiguous

Notes:

1 The CGS system is a coherent system of units that has centimeter, gram, and second as the base quantities for length, mass, and time respectively (SI units are preferred in engineering).

2 As a measure of plain angle, it is preferential to use decimals (e.g., 13.25°) rather than degrees, minutes, and seconds (this does not apply to cartography).

3 The British English spelling of liter is *litre.* It is better to use uppercase L for liters rather than lowercase l, which can be confused with the number one; however when combined with a prefix (e.g., ml), lowercase l is unambiguous and may be used.

4 The same symbol (abbreviation) is used for singular and plural (e.g., 1 lb, 2 lb; never write 6 mins or 187 kips). When the units are spelled out use the singular for values of one or less, but zero takes the plural form (e.g., zero inches, half inch, one inch, two inches).

5 Abbreviations of units of measurement are written without periods. This is always the case for SI units (which are called symbols); however, there are exceptions with non-SI units where, in non-technical writing, periods are used (e.g., in., sq. in., cu. in., ft., hr., min., sec.).

Units (SI)

- SI units (Système international d'unités) are the preferred system of units for most engineering publications. There are seven base units, two supplementary units and nineteen derived units (all listed below). There are also a few additional units that may be used with SI units (e.g., liter). Prefix symbols (listed below) are applied to the base units with the exception of mass where they are applied to the gram (hence, it would be incorrect to write μkg).
- SI unit symbols may be regarded as mathematical symbols (rather than ordinary abbreviations) in the sense that they can be manipulated in the same way as other mathematical symbols.
- The units are written lowercase when spelled out (this includes newton and pascal, and the other units named after famous people) with the exception of degree Celsius. However, both uppercase and lowercase are used to define the symbols.

Base SI units

Unit	Symbol	Quantity
ampere	A	electric current
candela	cd	luminous intensity
kelvin	K	thermodynamic temperature
kilogram	kg	mass
meter	m	length (see note 1)
mole	mol	amount of substance
second	s	time

Notes:

1. British English spelling: metre.
2. Use m^2 and m^3 for area and volume respectively and not sq m and cu m.
3. Common mistakes include writing candle for candela; Kelvin or degree kelvin for kelvin; °K for K; kilos or Kg for kg and sec for s.

Prefixes for SI units

Factor	Prefix	Symbol	Factor	Prefix	Symbol
10^{24}	yotta	Y	10^{-1}	deci	d
10^{21}	zetta	Z	10^{-2}	centi	c
10^{18}	exa	E	10^{-3}	milli	m
10^{15}	peta	P	10^{-6}	micro	µ
10^{12}	tera	T	10^{-9}	nano	n
10^{9}	giga	G	10^{-12}	pico	p
10^{6}	mega	M	10^{-15}	femto	f
10^{3}	kilo	k	10^{-18}	atto	a
10^{2}	hecto	h	10^{-21}	zepto	z
10^{1}	deca *or* deka	da	10^{-24}	yocto	y

Notes:

1. The prefix symbol is applied to the base unit (e.g., 1 ns, which equals 10^{-9} s), except for mass where it is applied to the gram (e.g., 1 mg, which equals 10^{-3} g or 10^{-6} kg).
2. Multiple prefixes (e.g., µµs) should not be used.
3. The prefix symbol should only be used with the base symbol and not when the unit is spelled out (e.g., write µs but not µsecond).

SI-derived units and SI supplementary units

Unit	Symbol	Equivalent	Quantity
becquerel	Bq	s^{-1}	activity of a radionuclide
coulomb	C	A·s	quantity of electricity, electric charge
degree Celsius	°C		temperature (see note 1)
farad	F	C/V *or* $C \cdot V^{-1}$	capacitance
gray	Gy	J/kg *or* $J \cdot kg^{-1}$	absorbed dose, absorbed dose index
henry	H	Wb/A *or* $Wb \cdot A^{-1}$	inductance
hertz	Hz	s^{-1}	frequency
joule	J	N·m	energy, work, quantity of heat
lumen	lm	cd·sr	luminous flux
lux	lx	lm/m^2 *or* $lm \cdot m^{-2}$	illuminance

SI-derived units and SI supplementary units (*continued*)

Unit	Symbol	Equivalent	Quantity
newton	N	$kg{\cdot}m/s^2$ *or* $kg{\cdot}m{\cdot}s^{-2}$	force
ohm	Ω	V/A *or* $V{\cdot}A^{-1}$	electric resistance
pascal	Pa	N/m^2 *or* $N{\cdot}m^{-2}$	pressure, stress
radian	rad		plane angle (see note 2)
siemens	S	A/V *or* $A{\cdot}V^{-1}$	electric conductance
sievert	Sv	J/kg *or* $J{\cdot}kg^{-1}$	dose equivalent, dose equivalent index
steradian	sr		solid angle(see note 2)
tesla	T	Wb/m^2 *or* $Wb{\cdot}m^{-2}$	magnetic flux density
volt	V	W/A *or* $W{\cdot}A^{-1}$	electric potential, potential difference
watt	W	J/s *or* $J{\cdot}s^{-1}$	power, radiant flux
weber	Wb	V·s	magnetic flux

Notes:

1 A change of 1 °C is equivalent to a change of 1 K, and 0 °C = 273.15 K.

2 Radian and steradian are supplementary SI units.

3 Periods are never used with SI symbols (they are not abbreviations).

4 The same symbol is used for both singular and plural (e.g., 1 V, 2 V).

5 See ISO 31:1992 for further details on SI quantities, units, and symbols.

URL

- A Uniform Resource Locator (URL) is an Internet address, which specifies the location of an object on the World Wide Web (WWW) or elsewhere on the Internet. It consists of the Internet protocol name, host name, and, optionally, other references such as the port, directory, and file name. The Internet protocol name (e.g., *HTTP*, *FTP*, *GOPHER*) is separated from the rest of the address by two forward slashes. In the case of HTTP, the protocol name is omitted in many addresses, as the browser assumes this, and adds the prefix *http://*. E-mail and bulletin board addresses are also URLs.

- HTML programs automatically underline hypertext links (i.e., pointers to other hypertext or Internet sites). As a consequence, the practice to underline URLs in other situations (e.g., when citing Internet references) is widespread. Additionally, some authors choose to enclose the address in angle brackets (this ensures that a punctuation mark following a URL is not mistaken for part of the address).
- Treat the URL as an *address* that requires the preposition *at* before it—for example: *Details regarding the CSE method for citing Internet sources can be found at* *www.councilscienceeditors.org/.*
- Be careful with the underscore mark (i.e., _) as the underlining may obliterate it, giving the appearance of a blank space. Some publishers do not underline URLs for this reason, choosing instead to write them in italics.
- It is advisable to include the trailing slash associated with a URL that does not have a file name (e.g., www.nasa.gov/) as this facilitates the server to return a website, rather than to cycle back to the browser, if the request was not for a valid "browseable" file.
- URLs can be very long and may extend onto a second line on a page. When writing a long URL do not introduce a hyphen or any other symbol to join the two parts; simply break the URL at a suitable location (e.g., after a slash) and continue on the next line.
- See also **Internet** and **Internet reference citation**.

Z

Zoology and biology terms

There is a standard format for writing the scientific (Latin) names of animals and plants—see **Scientific names (of animals and plants)**. For further details consult *Scientific Style and Format: The CSE Manual for Authors, Editors, and Publishers* (Rockefeller University Press, 2006), for example.

Index

A

B

C

Q

R

S

T